Biomedical Imaging Technology

Biomedical Imaging Technology

Signal Processing Strategies and Innovations

Edited by

Ayush Dogra
Chitkara University Institute of Engineering and Technology
Chitkara University, Punjab, India

Shalli Rani
Chitkara University Institute of Engineering and Technology
Chitkara University, Punjab, India

Ankita Sharma
Chitkara University Institute of Engineering and Technology
Chitkara University, Punjab, India

IEEE Press
WILEY

Published by John Wiley & Sons, Inc., Hoboken, New Jersey.

Library of Congress Cataloging-in-Publication Data Applied for:

Hardback ISBN: 9781394348053

Cover Design: Wiley
Cover Image: © sudok1/stock.adobe.com

Set in 9.5/12.5pt STIXTwoText by Straive, Chennai, India

This book is dedicated to the pioneers of biomedical imaging, whose vision and innovation laid the foundation for today's technological frontiers; the researchers, clinicians, and engineers, who continue to push boundaries and transform possibilities into realities; and most importantly, to the patients and communities worldwide, whose lives inspire and drive the pursuit of excellence in healthcare innovation.

Contents

List of Contributors

Suhaib Ahmed
Center for Applied AI
Model Institute of Engineering and Technology
Jammu
India

Azzah AlGhamdi
Computer Information Systems Department
College of Computer Science and Information Technology
Imam Abdalrhman Bin Faisal University
Khobar
Saudi Arabia

Vandana
Chitkara University Institute of Engineering and Technology
Chitkara University
Punjab
India

Puneet Bawa
School of Computer Science and Engineering
IILM University
Gurugram
India

Mohammed Wasim Bhatt
Model Institute of Engineering and Technology
Jammu, J&K
India

Gurbinder Singh Brar
School of Computer Science & Engineering
Lovely Professional University
Jalandhar
Punjab
India

Haewon Byeon
Workcare D-Health Lab
Convergence Department
Korea University of Technology and Education
Cheonan
South Korea

Ayush Dogra
Chitkara Institute of Engineering and Technology
Chitkara University
Punjab
India

Deepali Gupta
Chitkara University Institute of Engineering and Technology
Chitkara University
Punjab
India

Divya Gupta
Department of Computer Science and Engineering
Chandigarh University
Mohali
Punjab
India

Shubham Gupta
Center for Applied AI
Model Institute of Engineering and Technology
Jammu
India

Manju Jakhar
School of Pharmacy and Emerging Sciences
Baddi University of Emerging Sciences & Technology
Baddi
Himachal Pradesh
India

Jaspreet Kaur
Department of Computer Science and Engineering
Chandigarh University
Mohali
Punjab
India

Navneet Kaur
School of Computer Application
Lovely Professional University
Jalandhar
Punjab
India

Ashish Kashyap
Department of Pharmacology
Bhojia Dental College and Hospital
Nalagarh, Baddi
Himachal Pradesh
India

Ismail Keshta
Computer Science and Information Systems Department
College of Applied Sciences
AlMaarefa University
Riyadh
Saudi Arabia

Kamini Lamba
Chitkara Institute of Engineering and Technology
Chitkara University
Punjab
India

Sheenam Middha
Department of Computer Science and Engineering
Chandigarh University
Mohali
Punjab
India

Manisha Pathania
University Institute of Computing
Chandigarh University
Mohali
Punjab
India

Manisha Rajput
Chitkara University Institute of Engineering & Technology
Chitkara University
Punjab
India

Nidhi Rani
Chitkara College of Pharmacy
Chitkara University
Punjab
India

Shalli Rani
Chitkara Institute of Engineering and Technology
Chitkara University
Punjab
India

Chander Partap Singh
University Institute of Computing
Chandigarh University
Mohali
Punjab
India

Mohammad Shabaz
Department of Computer Engineering
Faculty of Engineering and Technology
Marwadi University Research Center
Marwadi University
Rajkot
India

Ankita Sharma
Chitkara Institute of Engineering and Technology
Chitkara University
Punjab
India

Chetna Sharma
Chitkara University Institute of Engineering and Technology
Chitkara University
Punjab
India

Thakur Gurjeet Singh
Chitkara College of Pharmacy
Chitkara University
Punjab
India

Mukesh Soni
Centre for Research Impact & Outcome
Chitkara University Institute of Engineering and Technology
Chitkara University
Rajpura
Punjab
India

Tanishq Soni
Chitkara University Institute of Engineering and Technology
Chitkara University
Punjab
India

Mudita Uppal
School of Computer Science and Engineering
IILM University
Greater Noida
UP
India

About the Editors

Dr. Ayush Dogra is currently serving as an Assistant Director-Research at the Chitkara University. Dr. Ayush Dogra is graduated with a Bachelor of Engineering and Technology from Guru Nanak Dev University, Amritsar, in 2011. He received his Master's degree in Engineering from Punjabi University, Patiala, in 2013 and Master's degree in Business Management (MBA) from IGNOU, Delhi (Central University) during 2015. He completed his doctoral degree in the Department of Electronics and Communications engineering from October 2015 to January 2019 from Panjab University. He worked as a Junior and Senior Research Fellow in the Department of Electronics and Communication Engineering, UIET, Panjab University, Chandigarh, and as a Post-Doctoral Fellow in the Department of Center of Biomedical Engineering, IIT Ropar. He has also worked as CSIR-Nehru Post-Doctoral Researcher in the Biomedical Applications Unit, CSIR-Central Scientific Instruments Organization, Chandigarh. He has published numerous papers (>180) in highly reputed SCI/SCIE/SCOPUS/PUBMED indexed journals and conferences. He is currently acting as editorial/reviewer work for various highly reputed SCI/SCIE and Scopus indexed journals. As of now, he has reviewed around 1300+ manuscripts for more than 140 journals and also handled 200+ manuscripts as an editor.

His work area focuses on devising a novel and innovative, market-oriented mechanism for medical imaging and Imaging applications. He is the recipient of several awards some of which are:

1) Dr. Paul Christian Lauterber, Young Scientist Award, World Congress on Cardiac Sciences, J. N. Tata Auditorium, Indian Institute of Science, Bengaluru

2) Top Peer Reviewer awarded for placing in the top 1% in engineering sciences/-Computer Science/Cross Field in Global Peer Review Awards Conducted by PUBLONS (Clarivate Analytics – Web of Science)
3) Research years (Research Award Winner for two consecutive years) from UIET, Panjab University, Chandigarh
4) Sushruta Most Popular Research Award by iERP DELHI.
5) Green Thinkerz Preeminent Researcher Award, Chandigarh

In addition, Dr. Ayush Dogra designed a series of optimized medical imaging algorithms for enhanced visualization of various ailments and presented it through many SCI/SCIE indexed research publications with high impact factors in a short span of time. He had significantly worked on medical image fusion and designed many algorithms for fusion of osseous and vascular images with high visual quality at much lower cost. Further, he is exploring the other types of medical modalities. During his doctoral he has worked on osseous and vascular-based fusion to enhance the details of vascular structures for detecting stenosis and aneurysm. He is also engaged in Low Dose Cardiac (Fluoroscopic and stent) image enhancement and CT-MR-based fusion for the detection of various abnormalities. OSSEOUS- and VASCULAR-based projects are being carried out with reputed medical colleges. Low-dose X-ray (Cardiac) image quality improvement is currently an on-going project with Medical Industry, Chandigarh, which holds significant importance for healthcare.

Dr. Shalli Rani (Director, Research) completed her post-doctoral research at Manchester Metropolitan University, UK, in June 2023. She is a Professor at the Chitkara University Institute of Engineering and Technology, Chitkara University, Punjab, India. She has 18+ years teaching experience. She received her MCA degree from Maharishi Dyanand University, Rohtak, in 2004, MTech degree in Computer Science from Janardan Rai Nagar Vidyapeeth University, Udaipur, in 2007, and PhD degree in Computer Applications from Punjab Technical University, Jalandhar, in 2017. Her main areas of interest and research are Wireless Sensor Networks, Underwater Sensor networks, Machine Learning, and Internet of Things. She has published, accepted, or presented more than 100+ papers in international journals and conferences (SCI+Scopus) and edited or authored five books with international publishers. She is serving as an Associate Editor of *IEEE Future Directions Letters*. She served as a Guest Editor for *IEEE Transaction on Industrial Informatics*, *Hindawi WCMC*, and *Elsevier IoT* journals. She has also served as a reviewer for many repudiated journals from IEEE, Springer, Elsevier, IET, Hindawi, and Wiley. She has worked on Big Data, Underwater Acoustic Sensors, and IoT to show the importance of WSN in IoT applications. She received a Young Scientist Award in February 2014 from Punjab Science Congress, and Lifetime Achievement Award and a Supervisor of the Year Award from Global Innovation and Excellence in 2021. Her work has gained global and reputed recognition, and she has been nominated as one of the top 2% of scientists in her field by Stanford University.

Dr. Ankita Sharma is an Assistant Professor at the Chitkara University Institute of Engineering and Technology, Chitkara University, Punjab, India. She received her BTech and MTech degrees from Ganpati Group of Institutions affiliated with Kurukshetra University. She has published, accepted, and presented more than 40+ papers in international journals and conferences (SCI+Scopus). Her research interests include Security domain in the Internet of Things using machine learning and deep learning.

Preface

Biomedical imaging has emerged as one of the most transformative domains in healthcare, enabling clinicians, researchers, and engineers to visualize, diagnose, and interpret complex biological processes with unprecedented precision. Over the past few decades, remarkable advancements in imaging modalities, computational methods, and signal processing strategies have revolutionized how we perceive health and disease, bridging the gap between technology and medicine.

The book *"Biomedical Imaging Technology: Signal Processing Strategies and Innovations"* brings together leading perspectives and contributions from researchers and practitioners to explore the evolving landscape of biomedical imaging. It is structured to provide a comprehensive understanding, beginning with the historical evolution of imaging technologies and advancing toward state-of-the-art innovations shaping the future.

The chapters cover a wide spectrum of themes from deep learning techniques and advanced image reconstruction methods to multimodal and molecular imaging, fluorescence-based approaches, and emerging diagnostic protocols. Special emphasis is placed on integrative frameworks for image analysis, therapeutic interventions guided by imaging, and the challenges faced in transitioning next-generation imaging technologies into clinical practice.

This volume is designed not only to highlight technological breakthroughs but also to reflect on their clinical applications and broader implications in improving patient outcomes. The diversity of contributions from global authors ensures a multi-dimensional perspective, making this work a valuable resource for researchers, academicians, clinicians, and students who aspire to understand and contribute to the domain of biomedical imaging.

Acknowledgments

First and foremost, we would like to thank the Almighty and our family members for standing beside us throughout our research career so far and editing this book. Our families have been our motivation for continuing to improve our knowledge and move forward in our career. They are our lifetime achievement, and we dedicate this book to them.

Here, we are also thankful to the team who helped us in every aspect in the preparation of this book. Their prompt response and care to the literature and contribution were remarkable and sensational. They are great and responsible researchers indeed.

We would also like to thank each and every one behind the completion of book especially colleagues, who helped a lot, and without their quick and efficient efforts, it was not possible to get our book published. They motivated and gave us this platform to go ahead.

Last but not least, our gratitude is due to our management of **Chitkara University, Punjab**, for their trust in us and for making us confident that we can put efforts to get success in each and every field.

1

Historical Evolution and Technological Advancements in Biomedical Imaging

Shubham Gupta and Suhaib Ahmed

Center for Applied AI, Model Institute of Engineering and Technology, Jammu, India

1.1 Introduction

The realm of biomedical imaging is one of the keys to modern healthcare and its multidisciplinary approach, which has transformed diagnosis, treatment, and disease monitoring. Recent developments in signal processing have opened new ways of improving image quality, extracting valuable information, and performing real-time analysis [1]. These technologies are fusing innovative algorithms and machine learning (ML) to expand the horizons of accuracy and efficiency. The chapter discusses the most recent signal processing techniques that are fueling advances in the field of biomedical imaging, with emphasis on their innovative impact on clinical and research perspectives.

1.1.1 Role of Biomedical Imaging in Modern Medicine

Biomedical imaging is the foundation of contemporary medicine and is an essential approach for the diagnosis, treatment, monitoring, and management of different diseases. It is a major medical breakthrough that can help clinicians visualize internal structures non-invasively and enables accurate and rapid clinical decision making. Imaging modalities (i.e. X-ray, computed tomography (CT), magnetic resonance imaging (MRI), ultrasound (USG)) are ubiquitous and irreplaceable tools of clinical practice worldwide.

For instance, in oncology, **MRI offers higher sensitivity for detecting soft-tissue contrast in brain tumors**, while positron emission tomography **(PET) imaging is preferred for detecting metabolic activity** related to cancer spread. In cardiovascular imaging, **echocardiography and cardiac MRI** are

Biomedical Imaging Technology: Signal Processing Strategies and Innovations, First Edition.
Edited by Ayush Dogra, Shalli Rani, and Ankita Sharma.

Table 1.1 Biomedical imaging in modern medicine.

Aspect	Description	Impact on modern medicine
Early detection and diagnosis	Advanced imaging modalities like MRI, CT, PET, and ultrasound are used to identify abnormalities	Early and accurate detection of diseases, such as cancer, cardiovascular conditions, and neurological disorders, leads to improved prognosis and timely intervention
Treatment planning	Provides detailed anatomical and functional insights for planning surgeries or therapies	Enables precise localization of tumors, identification of surgical pathways, and assessment of treatment targets, reducing risks and improving treatment outcomes
Real-time monitoring	Facilitates intraoperative imaging and real-time guidance during surgeries or minimally invasive procedures	Enhances surgical precision and ensures effective execution of interventions, such as tumor resections or catheter placements
Personalized medicine	Integrates patient-specific data to customize imaging protocols and treatment plans	Improves diagnostic accuracy and patient outcomes by tailoring imaging techniques and therapies to individual needs
Disease progression	Monitoring tracks changes in disease states using longitudinal imaging studies	Enables clinicians to evaluate the effectiveness of treatments and adjust therapeutic strategies as needed

often combined to assess both function and structure. The choice of modality impacts diagnosis and prognosis and is increasingly tailored using **clinical decision support systems (CDSS)** driven by imaging data (Table 1.1).

Biomedical imaging serves other purposes beyond diagnostics. It plays a basic part in influencing therapeutic interventions, including picture-guided medical procedures and, for example, radiotherapy [2]. Interventional radiology uses imaging to apply treatments as carefully as possible to tumors or other targets to minimize damage to nearby tissue. In addition, imaging plays a critical role in assessing therapeutic efficacy, such as monitoring the reduction in tumor volume after chemotherapy or the status of wound healing.

It would provide new knowledge of biological kinetics at the molecular and cellular scales for medical research in biomedical imaging. in vivo techniques, including functional imaging (fMRI, PET), enable researchers to investigate dynamic brain activation, metabolic processes, and more general disease mechanisms. This has pioneered new territories in neuroscience, oncology, cardiology, and beyond. Furthermore, integrating imaging with artificial intelligence (AI) and ML has been essential in precision medicine, enabling the use of individualized treatment strategies based on a given patient's unique imaging findings.

1.1.2 Historical Context of Imaging Techniques

Over the last few centuries, biomedical imaging has evolved to what we see today, an intermediate between several critical discoveries and breakthroughs that have redefined our perspective of the human body. It all began with the discovery of X-rays by Wilhelm Roentgen back in 1895. With this breakthrough, doctors could view the human body for the first time and provide the basis of diagnostic radiology. X-ray imaging rapidly established itself as a staple of diagnosis, changing the diagnosis of all types of fractures, infections, pulmonary diseases, etc.

Several imaging modalities were invented during the mid-20th century. Accordingly, in the 1950s, USG imaging was developed as a safe tool to assess soft tissue and organ systems by taking inspiration from sonar imaging [3]. This was especially groundbreaking in the field of obstetrics, where fetal development can now be monitored safely. Concurrently, the origination of nuclear imaging techniques like single-photon emission computed tomography (SPECT) and PET made possible the visualization of physiologic processes and functional imaging to complement anatomical imaging.

Another advancement came with CT, invented in the 1970s by Godfrey Hounsfield and Allan Cormack as shown in Figure 1.1. CT, on the other hand, combined X-ray exposure (CT stands for computed tomography) with advanced computer algorithms to generate images of the body in cross-section, with much more detail than was possible through traditional

X-ray scans. Around then, MRI came along, providing high-contrast imaging of soft tissues. Based on the principles of nuclear magnetic resonance, MRI quickly became utilized in the fields of neurology, orthopedics, and oncology. Optical imaging methods, including fluorescence microscopy and confocal microscopy, have also diversified and expanded the possibilities for biomedical imaging. Those methods allowed visualization with cellular and subcellular resolution, which has resulted in major advances in molecular biology and pathology.

1.1.3 Objectives and Scope of the Chapter

This chapter aims to serve as a comprehensive synthesis of traditional, modern, and emerging signal processing methods used in biomedical imaging, grounded in both theoretical concepts and practical applications. It is particularly

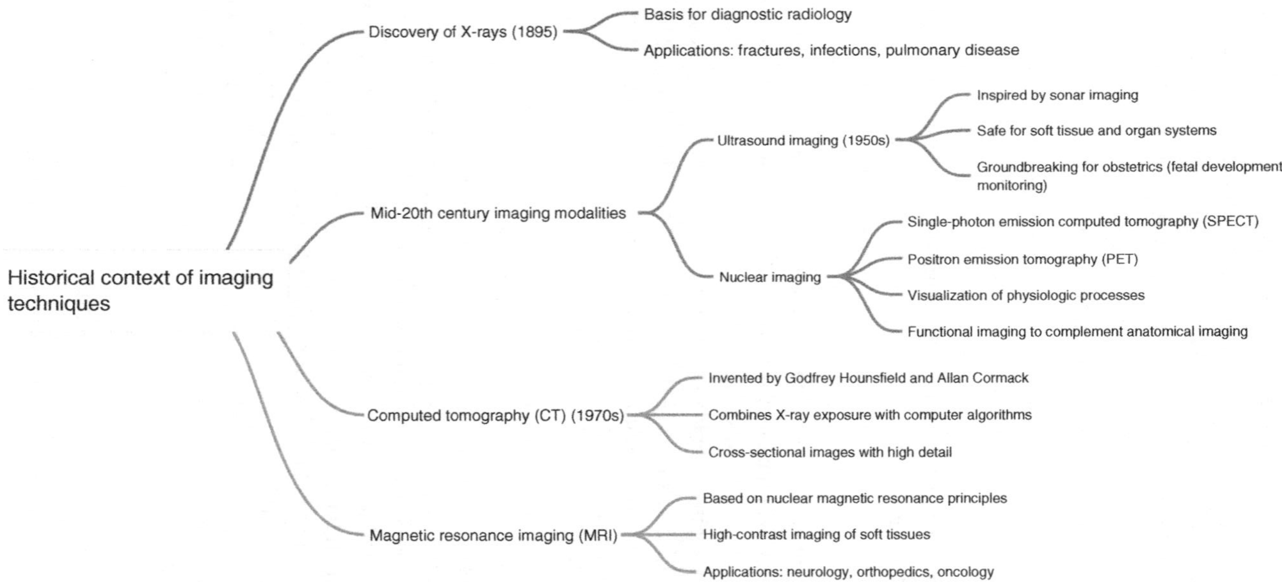

Figure 1.1 Historical context of imaging techniques.

relevant for clinicians, researchers, and policymakers seeking to understand the multidisciplinary evolution of biomedical imaging. By integrating current research, AI-driven innovation, and challenges related to regulation and accessibility, we aim to contribute to evidence-based and ethically responsible advancements in healthcare.

1.2 Early Milestones in Biomedical Imaging

Following centuries of development and discovery, biomedical imaging has played a crucial role in the progress of modern medicine. Before the availability of imaging technologies, physicians depended on narrow diagnostic instruments and observational methods. The path from rudimentary anatomical studies to advanced nuclear imaging captures humanity's unflagging desire to comprehend the human body and its diseases [4]. This chapter covers the early days of biomedical imaging, along with the pre-imaging period, the epoch-making discovery of X-ray, followed by radioisotope imaging as shown in Figure 1.2.

1.2.1 Pre-Imaging Era: Anatomy and Physical Diagnosis

Earlier, physicians were dependent on anatomy and physical diagnosis. Physicians had to rely on external telltales, probing with their fingers or with the stethoscope to identify internal infirmities. Ancient Egypt and Greece are two examples of classical civilizations that made substantial contributions to early anatomical knowledge, with their people having studied human remains extensively. Without imaging tools, medicine was built on the pioneering work of anatomists such as Hippocrates and Galen. Their scale of bodily functions and pathologies were primitive by later standards, but they emphasized the crucial role of observation. However, these methods had a limited scope. While invaluable, autopsies were often hampered by cultural and religious taboos, limiting anatomical investigation.

From the Renaissance onward, systematic dissections resulted in major improvements in anatomical knowledge. The Great Andreas Vesalius left his mark on medical education in the second half of the 16th century when he published a series of detailed anatomical illustrations in the *De Humani Corporis Fabrica* (1543), a publication with which he is often associated as essential for the establishment of modern anatomy [5]. These were real studies, but they were static models, without the provision to visualize dynamic physiological processes. In the 18th and 19th centuries, physical diagnostic methods began to be developed with, for example, René Laennec's invention of the stethoscope in 1816. These tools amplified the physician's capacity to directly sample internal conditions. However, the inability to access the living body was a major impediment to accurate diagnosis and treatment.

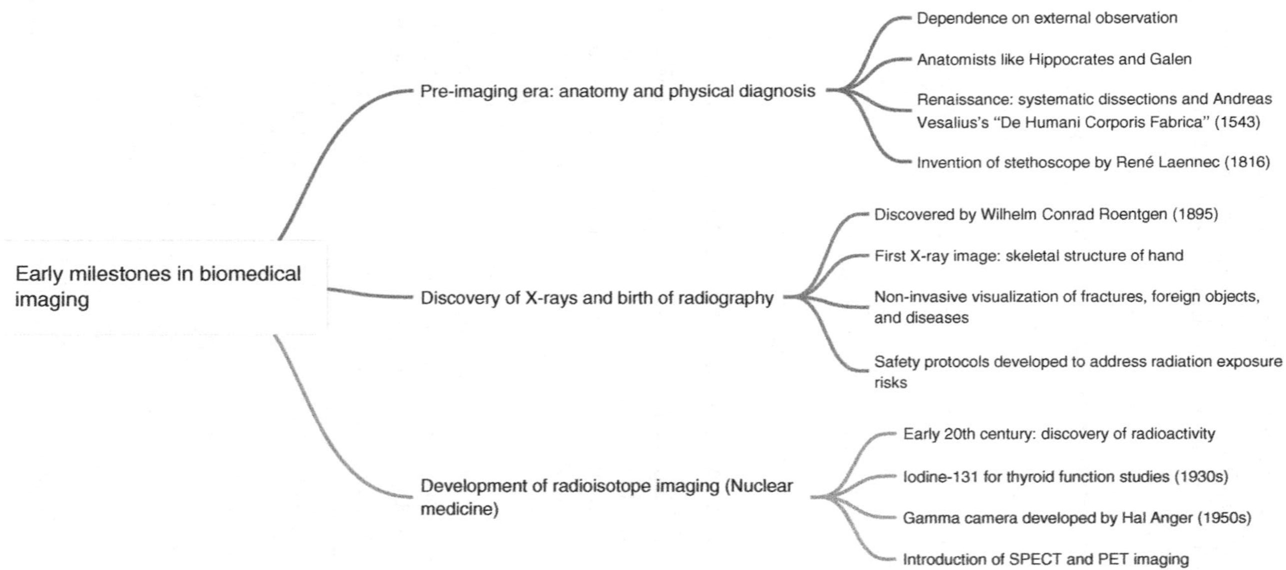

Figure 1.2 Early milestones in biomedical imaging.

1.2.2 Discovery of X-Rays and Birth of Radiography

The discovery of X-rays by Wilhelm Conrad Roentgen in 1895 ushered in a transformative era in medicine. Roentgen's chance finding of an unexplored type of radiation during experiments with cathode rays changed the field of diagnostics. His first X-ray picture of human anatomy showed his wife's hand and the skeletal structure of her hand, illustrating the possibilities of this invention.

X-rays provided the first non-invasive way to visualize the inside of the human body, quickly becoming indispensable for diagnosing fractures, spotting foreign objects, and identifying diseases such as tuberculosis. Crude early radiographs required long exposure times and resulted in poor image quality [6]. Despite having these limitations, X-ray technology was applied by the medical community in no time as it was found to have great potential.

Radiographic methodology progressed rapidly over the next several years. Developments included high-powered screens, which allowed for shorter exposure times, and contrast agents allowed for visualization of soft tissue. Specialized X-ray procedures such as fluoroscopy and angiography in the mid-20th century expanded the diagnostic range of radiography.

With the discovery of X-rays, interdisciplinary studies commenced that led to research and application of radiation therapy for the treatment of cancer. This dual role of the X-ray in diagnostics and therapy highlighted its importance in contemporary medicine [7]. However, the detrimental health effects associated with prolonged radiation exposure, underestimated at first, spurred the development of safety protocols and protective measures.

1.2.3 Development of Radioisotope Imaging (Nuclear Medicine)

The development of radioisotope imaging added a third dimension to biomedical imaging, as it allowed visualization of physiological functions in addition to anatomical structures. Nuclear medicine was established at the beginning of the 20th century when radioactivity was discovered. These discoveries sparked interest in the use of radioactive materials for medical applications.

Clinical utility for radioactive isotopes began in the 1930s with the use of iodine-131 for thyroid function studies. This was the beginning of functional imaging where body distribution of radioisotopes could provide information about metabolic and pathophysiologic processes. Unlike X-rays that showed static structural flaws, radioisotope imaging revealed organ motion functions and dynamic features that were valuable in identifying and assessing diseases like hyperthyroidism and cardiac ischemia. Nuclear medicine was revolutionized with the introduction of specialized imaging apparatus, particularly the gamma camera designed by Hal Anger in the 1950s. This was followed by the incorporation of gamma cameras into the detection process, which then allowed the visualization of the distribution of the radiopharmaceuticals in real time and more confident diagnosis [8]. Later developments in nuclear imaging offered increased spatial and temporal resolution with the advent of SPECT and then PET.

PET imaging became important in oncology, neurology, and cardiology since it was able to visualize metabolic processes and molecular interactions at a cellular level. Spotting cancerous tissues like never before, the advent of the incredible Fluorodeoxyglucose (FDG) as a radiolabeled analogy of glucose completely transformed how cancer diagnosis and staging was done. Nuclear medicine offered many advantages, but it also had its restrictions, such as the safety of radiation, complicated radioisotope operations, and being expensive. Efforts to address these obstacles include developing safer dynamic imaging approaches on a blood or peanut scale.

The early milestones in the area of biomedical imaging highlight the transformative influence the field has had on medicine [9]. The evolution of medical imaging, from the nascent anatomical dissections before imaging technology to the revolutionary X-ray to the functional information afforded by radioisotope imaging, set the stage for the varied diagnostic arsenal we now possess. Such advances not only transformed our knowledge of the human body but also provided the basis for future breakthroughs in imaging science.

The foundational technologies discussed in this section laid the groundwork for the current era of digital and computational imaging. Building upon these early modalities, Section 1.3 delves into the core signal processing strategies that enhance image clarity, diagnostic precision, and real- time analysis in modern biomedical imaging systems.

1.3 Signal Processing Strategies in Biomedical Imaging

Signal processing approaches are important in biomedical imaging to convert raw data into diagnostic, high-resolution images that assist with diagnosis and therapy. The entire process, from raw signal acquisition through image processing and reconstruction, is optimized to perform accurately, repeatably, and efficiently [10]. We provide a more in-depth discussion of some of the critical signal acquisition, preprocessing, reconstruction, feature extraction, and real-time processing innovations that have facilitated the advancement of the field.

1.3.1 Data Acquisition and Preprocessing

Data acquisition is the foundation of biomedical imaging. It collects raw signals from sensors, including those placed inside different organs, through the Imaging Calibration and Acquisition Noise Suppression (ICANS) device. Raw signals are another name for the same output signals, but they are frequently polluted with noise from different sources like thermal noise, deformed signals of electronics, and biological motion signals like patient breathing or heart rate.

Noise reduction is an essential preprocessing step since noise can hide important features of biomedical images. For this purpose, adaptive filtering techniques have become an effective tool. As an example, a statistical technique called Wiener filtering will seek to minimize mean square error by adapting to the underlying signal properties. This method has shown success in MRI and CT imaging, where high levels of noise in images can obscure smaller anatomical features. Another adaptive method, Kalman filtering, has excellent usefulness in real-time applications such as within USG imaging when new signals need to be used quickly [11]. Indeed, Kalman filters are effective for tracking dynamic changes over time by estimating the underlying state of a system from potentially noisy observations, which makes them widely applicable in contexts such as imaging with motion.

Signal calibration is another key component of preprocessing. This provides a uniform and precise measurement system by correcting detector bias, normalizing signal intensity, and compensating as per the system's non-linearity. One common issue in many modalities, like X-ray and optical imaging, is the ambient variations in detector sensitivity; to rectify that, flat-field correction techniques are being used. On the other hand, phantom-based calibration approaches rely on reference objects with known physical characteristics to assess system performance. The above methods are particularly important to maintain reproducibility and accuracy in research and clinical settings.

1.3.2 Image Reconstruction Algorithms

The process of reducing raw signal data into interpretable imagery and image reconstruction allows clinicians and researchers to visualize internal structures and functions. Developments in computational methods have greatly improved image reconstruction quality and speed across multiple modalities.

For decades, the Fourier transform (FT) has been an essential building block of image reconstruction procedures, particularly in the field of MRI and CT imaging. Such an operation allows for local manipulation (editability and high resolution) and local visualization of periodic structures by transforming the data in the spatial domain to the frequency domain. MRI, for instance, employs the k-space representation of signals, and the FT is crucial in transforming frequency information into spatial coordinates [12]. This approach provides high-quality images on hardware with limited computational capabilities.

The wavelet transform (WT) suggests a quite flexible method due to the possibility of multiresolution analysis of the signals. Conversely, the WT decomposes signals as a sum of components for each scale; a single scale of the FT exists which causes the loss of localization in both domains. This characteristic becomes an advantage for biomedical analysis that needs precise identification, such as classifying minor variations of tissues from USG or CT images. The WT is widely used for image denoising such as retaining image structure while preserving image accuracy.

Over the last few years, compressed sensing (CS) has transformed image reconstruction by allowing high-quality images to be reconstructed from dramatically fewer data samples. Magnetic Resonance Adaptive Compressed Sensing (MRACS) reconstructs images by leveraging the characteristic sparseness of biomedical signals in some specific domains (e.g. wavelet or gradient domains), allowing minimal information loss during reconstruction. MRI has been revolutionary, reducing scan times drastically and thus improving patient comfort and increasing throughput in clinical workflows. In CS, algorithms such as LASSO (Least Absolute Shrinkage and Selection Operator) are often applied to obtain sparse signals with low complexity, showcasing the impact of computational strategies on biomedical imaging.

1.3.3 Feature Extraction and Enhancement

Post image reconstruction, processes are centered on feature extraction and enhancement, which play a major role in finding and enhancing the regions of interest in the images. These are essential steps for a correct diagnosis and quantification.

Feature extraction is a fundamental step, such as edge detection, which is used to find the contours present in an image or the regions where two different structures or tissues meet. Edge detection techniques like Sobel and Canny operator locate edges based on having large gradients of intensities among pixels with only a few pixels differing, providing a high level of accuracy. For instance, in CT image processing, edge detection can isolate bony structures from the motion areas of soft tissues for a more detailed anatomical study [13]. Such advanced gradient-based methods are specifically for coping with the inferiority due to the noisy image and preserving important edges of the region.

Texture analysis focuses on quantifying surface features, such as smoothness, roughness, and asperity, that often are indicative of underlying vasculature. Texture variations in the regions, for example, allow distinguishing malignant from benign tumors in images obtained by MRI or USG imaging. Many texture analysis algorithms such as gray-level co-occurrence matrix (GLCM) and local binary patterns (LBP) are being used. These methods give a detailed statistical description of the image texture, which is useful for pattern recognition, and can simplify automated diagnosis systems.

Adaptive histogram equalization (AHE) redistributes today's intensity values over a smaller region in the image to increase contrast. Unlike traditional histogram equalization, which uses a uniform transformation for all pixels, AHE maintains edge details and enhances visibility in low- contrast regions [14]. Cumulatively, these enabled practitioners of these imaging modalities to identify changes in tissue density, which is incredibly useful for imaging modalities such as mammography.

1.3.4 Real-Time Processing Strategies

The need for real-time processing arises in biomedical imaging applications where immediate visualization and analysis are needed, such as in interventional radiology, emergency diagnostics, and surgical guidance. The hardware and algorithmic advances have brought about systems that can meet these performance goals.

GPUs (graphics processing units) are now a fundamental basis for real-time biomedical imaging through parallel processing. Because of their highly parallel architecture, GPUs can process large volumes of data in parallel, which makes them suitable for computationally intensive workloads such as 3D image reconstruction. For instance, in CT imaging, GPU acceleration greatly decreases reconstruction times, allowing for real-time representation of dynamic processes (blood flow). Similarly, it is in the rendering that GPUs also enable high-frame-rate rendering in USG imaging, allowing clinicians to evaluate the movement of organs in real time.

Additionally, to improve low-latency processing, specialized architectures to reduce the computational delay are developed. Such architectures typically combine pipeline processing with hardware accelerators (HA) that include FPGAs (field-programmable gate arrays) or ASICs (application-specific integrated circuits). These systems are most useful in dynamic imaging procedures such as cardiac MRI or fluoroscopy, where even slight delays can affect the diagnosis accuracy or procedure results [15]. They combine state-of-the-art algorithms with improved hardware to facilitate smooth operations in vital clinical situations.

The above strategies are considered the pillars of contemporary biomedical imaging for the translation of raw signals into clinically relevant, high-quality images for decision making. The edge of biomedical imaging capabilities has been challenged by signal processing through advancements in noise suppression, image reconstruction, feature extraction, and real-time processing. At the forefront of this transformation, these strategies are likely to shape the future of the healthcare industry for years to come.

1.4 Innovations in Signal Processing for Biomedical Imaging

Medical imaging in the 21st century signals how signal processing has transformed biomedical imaging, achieving the holy grail of precision, speed, and diagnostic sensitivity; new techniques in ML, quantum computing, multimodal integration, and dedicated hardware are all facilitating this progress. We briefly review some of the recent breakthroughs in signal processing and their groundbreaking impact on biomedical imaging.

1.4.1 Machine Learning and AI-Driven Techniques

ML/AI-Empowered Biomed Imaging—These have novel capabilities to automate intricate tasks and boost diagnostic performance and image quality. Deep learning (DL) is at the fundamental core of some recent advances in AI in the area of biomedical imaging. Neural networks, in particular convolutional neural networks (CNNs), are widely used for automatic diagnosis. CNNs are useful for image analysis, so for example, we can use them for tumor detection, organ segmentation, and disease classification. DL has demonstrated state-of-the-art performance in several tasks, such as the detection of lung nodules in chest CT and the classification of breast lesions in mammograms [16]. These models not only reduce the burden on the radiologists but also remove the element of human error, and enable repeatable and objective analysis in the process. DL also tackled issues related to relatively small amounts of annotated data available in most medical imaging applications by novel transfer learning methods and new architectures for pre-trained (typically ImageNet) models.

GANs, which stand for generative adversarial networks, are used for image enhancement. A GAN consists of a generator, a discriminator, and a pair of neural networks that can generate high-quality synthetic data. They are developed and tuned in opposite ways. For biomedical imaging, GANs are used to improve resolution, remove noise, and interpolate missing data. For instance, GANs have been applied to high-resolution MRI reconstruction from low-resolution scans, aiding better utility of MRI scans for diagnostic purposes without incurring any increase in acquisition times. Also, GAN-based methods have been used to generate training data for rare medical diseases, thus countering the data scarcity problem during the training of an AI model.

1.4.2 Quantum Signal Processing in Imaging

Signal Processing in the Quantum Computing Era: Unraveling the Potential to Transform Healthcare—Quantum optics uses atomic states to manipulate classical light in a way, which can be harnessed to generate high-speed data transfers, where the quantum nature of light demands quantum signal processing to enhance imaging abilities compared to classical approaches.

High-resolution imaging is one of the main areas of quantum signal processing applications. Quantum algorithms, like the quantum Fourier transform (QFT), can help accelerate the reconstruction of imaging data, thereby lowering the time needed for modalities such as MRI and CT. Similarly, at the quantum limit of sensitivity, quantum sensors have been shown to detect weak signals more effectively. Such capability is especially invaluable in applications such as optical coherence tomography (OCT) and molecular imaging where the acquisition of minute structures is crucial.

Potential development of image registration and segmentation using quantum techniques – As an illustration, hybrid quantum–classical frameworks apply quantum optimization in the alignment and feature extraction stages of multimodal imaging, which can enhance the accuracy of fused imagery [17]. Quantum signal processing is still in its infancy; however, it is a groundbreaking paradigm shift that may unlock a level of complexity in biomedical imaging systems that few have imagined available in the next few decades.

1.4.3 Multimodal Imaging and Data Fusion

Multimodal imaging is an increasingly appealing option that takes advantage of the complementary nature of different imaging modalities to describe biological structures and processes more completely. This is where signal processing comes into play, controlling the whole integration and fusion of multimodal data and enabling their smooth combination and understanding.

Multiple modality integration, in other words, PET–CT, MRI–PET, and CT–USG, allows the anatomy, functional, and molecular information to be viewed in one dataset. A combination of techniques from the world of signal processing, including mutual information-based registration and wavelet-based fusion, are leveraged to register and fuse image data collected from disparate modalities [18]. For example, in PET–CT imaging, the detailed structure provided by the CT scans is combined with the metabolic process images provided by the PET scans, and this allows precise localization of tumors for accurate planning of the subsequent radiation therapy.

Different AI-based algorithms also help to easily extract complementary features from different types of datasets, thus improving the multimodal data fusion process. DL models can learn complex relations between modalities, producing fused images while preserving relevant data of interest from both sources. This ability to layer data is important in clinical applications like brain imaging, where functional MRI (fMRI) can be combined with diffusion tensor imaging (DTI) to provide information on neural activity and structural connectivity, respectively. Multimodal imaging utilizes sophisticated signal-processing techniques for superior diagnostic precision and broader biomedical usage.

1.4.4 Emerging Trends in Signal Processing Hardware

Hardware technologies have evolved rapidly in the past decades, fostering several developments in biomedical signal processing. Complex hardware architectures are on the rise to match the demand for real-time processing, high-density image capture, and low power.

AI accelerators (GPUs and TPUs [tensor processing units]) are increasingly part of biomedical imaging workflows. Designed for high parallelization, these processors efficiently and quickly run complex algorithms, an essential requirement for tasks such as 3D image reconstruction and DL inference. Some examples include the utilization of GPUs in MRI and CT systems to process large datasets in real time, and the implementation of TPUs for the performance acceleration of AI models on diagnostics imaging applications.

Beyond generalist accelerators, ASICs are being pursued for specific imaging purposes. Joint- sourced ASICs enable a high level of customization, making them suitable for energy-efficient processing in portable or wearable imaging devices. As an example, ASICs enable handheld USG devices to offer real-time imaging capabilities in remote or resource-limited environments.

Interestingly, this formation ushers in some trends; for one in hardware, utilizing neuromorphic computing in signal processing is also being seen. Neuromorphic chips will replicate the structure and functionality of the human brain during operation, providing low-power, high-speed complex data processing. These chips have the potential for use in applications involving real-time pattern recognition and anomaly detection in biomedical imaging. Together, state-of-the-art algorithms and hardware Xenon systems are enabling next-gen biomedical imaging systems.

Emerging signal processing approaches are revolutionizing biomedical imaging, with the help of advancements in quantum computing, multimodal integration, and novel hardware advancements. These advances are markedly enhancing imaging quality and efficiency and paving the way for applications that previously appeared unattainable [19]. As such, their integration into biomedical imaging systems promises to improve diagnostic accuracy, patient care, and healthcare quality as these technologies continue to evolve.

1.5 Case Studies

The best way to understand how to use advanced signal processing techniques is through case studies in practice. Herein, we have targeted principal innovations in MRI, USG, and hybrid imaging modalities, and outlined how these innovations overcame diagnostic barriers, reduced procedural impediments, and facilitated improved clinical outcomes.

1.5.1 Innovations in Signal Processing for MRI

MRI is a beloved workhorse of modern diagnostic imaging and is prized for its non-invasive nature, and its ability to visualize soft tissues so extraordinarily well.

Classic MRI acquisition methods, however, are limited by long scan times and motion artifacts, which pose a problem in dynamic imaging contexts like cardiac or functional MRI. Signal processing techniques, especially CS, have effectively overcome the shortcomings.

By taking advantage of the trapping nature of MRI signals in some variables, compressed imaging reconstructs high-fidelity images from the data samples significantly less than the Nyquist–Shannon sampling theorem. This has transformed dynamic imaging, where temporal resolution is paramount. For instance, in cardiac MRI, CS facilitates the acquisition of hundreds of individual frames per second, which lets clinicians view the motion of the heart in lived time and without sacrificing spatial resolution. Again, the iterative reconstruction algorithms used in CS (such as total variation minimization) preserve sharp edges and small details in the images.

The clinical impact of CS goes far beyond efficiency. The reduced scan times translate to less patient discomfort and a decreased need for sedation, especially in pediatrics or claustrophobic patients. Moreover, the decreased data acquisition demands offset motion artifacts, resulting in clearer images that improve diagnostic accuracy [20]. This remarkable contribution combined with existing advanced techniques such as parallel imaging and DL-based reconstruction have ushered in a new era of reliability and applicability of MRI to research and clinical practice.

1.5.2 Deep Learning in Ultrasound Imaging

USG imaging is one of the most widely used medical technologies for diagnostic imaging which is cost effective, portable, and provides real-time data. However, USG images still suffer from speckle noise and low contrast that might obscure critical details and diagnoses. DL algorithms have proved to be very effective for image denoising and interpretation, minimizing the difficulties posed by traditional methods.

The concept of DL in USG image enhancement derives from the use of network algorithms that are trained to reduce noise and increase contrast, such as CNNs. DL-based denoising models trained on diverse sources are a promising alternative for speckle denoising requests without altering the common form of anatomical structures. Likewise, super-resolution networks are also a great option for denoising, USG image resolution enhancement, and localization of small lesions or subtle tissue changes.

In addition to improving images, DL has changed USG data interpretation by automating organ segmentation and disease identification. A notable instance includes obstetric imaging, where DL models precisely delineate fetal structures, facilitating earlier abnormality identification. Consider breast USG, where CNNs have been used to classify lesions as benign or malignant with diagnostic accuracy on par with that of the most experienced radiologists. Additionally, real-time USG systems with integrated DL have allowed for the evolution of AI-assisted devices that give immediate feedback to clinicians during USG-based imaging.

DL has greatly broadened the diagnostic scope of this modality by enhancing the quality of USG imaging and by speeding up their interpretation, making USG a more widely available agent of diagnosis.

1.5.3 Hybrid Imaging Modalities

PET–CT is a hybrid imaging technique that combines the spatial information from anatomical imaging with the functional information from nuclear medicine, providing a more comprehensive view of the diagnostic findings. PET is based on the tissue's chemistry and metabolism, and CT provides very high-resolution anatomy. These modalities are integrated using standard signal processing methods such as fusion and image registration.

PET–CT imaging integration approaches, including signal fusion, registration, and combination of the individual modality data, create a single image together. Typically, PET images are registered to CT images using mutual information-based registration algorithms to align PET and CT images so that the former matches anatomical information from CT images. This ensures that MET values are preserved from PET and very fine high-altitude anatomical details from CT images are kept, for wavelet-based fusion techniques. PET–CT allows the use of imaging from both modalities, increasing the diagnostic value of the method applied in the diagnosis and monitoring of many cancers (detection and localization of tumors, assessing response to therapy, estimating the amount of radiation needed), e.g. narrow AI for signal processing in hybrid imaging. Moreover, several DL models were developed for directly generating CT-like anatomical images from PET data, allowing some applications to remove dual acquisitions. These models then speeded up the imaging process and lowered radiation dosage which tackle the most important shortcomings of the standard PET–CT systems [21], just as recent advancements in signal processing formed the foundation for the success of hybrid imaging modalities. These systems offer a holistic view of intricate diseases by combining the strengths of diversified approaches, thereby improving diagnostic precision, and aiding in personalized therapeutic plans.

The above case studies demonstrate how the advances of evolution and methodology in signal processing are helping solve the riddles of multimanifold biomedical imaging successfully. This should involve advanced imaging techniques such as MR and US, including concepts of CS, DL, and photoacoustic signal fusion in hybrid imaging which have enhanced diagnostic precision, comportment, and clinical applicability. Interestingly, these latest profiles are not only proving beneficial for citizens to maintain health, but there is also scope for greater improvement in these domains.

1.6 Challenges and Future Directions

While there are exciting innovations, at the same time, there are real challenges that need to be solved for these technologies to reach their widespread potential. Not only that, the field is also on the cusp of critical growth, as new trends emerge to alter how imaging systems work and serve healthcare needs. In this section, we present the most critical issues and future prospects of biomedical signal processing.

1.6.1 Ethical and Regulatory Concerns

Applications of advanced digital signal processing such as those involving AI and ML create significant ethical and regulatory concerns. The first area of concern has to do with the explainability and transparency of popular AI models in biomedical imaging. DL algorithms are a form of ML that operates as a black box, user-determined inputs and outputs allowing for little understanding of why a particular diagnostic conclusion was reached. Because physicians are less likely to trust systems they do not understand, there is concern that such black boxes may inhibit clinical adoption.

The third ethical challenge pertains to data privacy and security. Biomedical imaging systems that handle sensitive patient data are hot targets for cyber-crime. Increased reliance on cloud-based and Internet of Things (IoT) networks for imaging systems may have marred previous assumptions that security and confidentiality of imaging data were minimal, whereby confidentiality and integrity of imaging data are equally critical. This ensures protection of patient data through regulatory compliances such as General Data Protection and Transparency (GDPT) and Health Insurance Portability and Accountability Act (HIPAA).

As such, regulatory frameworks will also need to evolve to address challenges related to the validation and approval of AI-based imaging systems. Approval processes designed for deterministic behavior in hardware and software likely will not be agile enough for AI models. However, these technologies must be implemented safely and effectively in the clinic, which requires strong standards for the evaluation, testing, and certification of these technologies.

For instance, the U.S. FDA has introduced "Software as a Medical Device (SaMD)" framework, which emphasizes model interpretability and post-market monitoring. The EU AI Act proposes risk classifications that could restrict black-box AI in high-risk sectors like radiology. These regulatory developments highlight the need for explainable AI (XAI) integration in biomedical imaging tools.

1.6.2 Scalability and Cost Effectiveness of Signal Processing Techniques

Though utilizing advanced signal processing techniques has shown great promise, variability in scalability and cost often inhibits their implementation. Many advanced approaches, e.g. DL-based image reconstruction or quantum signal processing, demand considerable computational capacities, which are unamenable in resource-limited environments.

These could be, for example, real-time imaging systems that use GPUs or TPUs, which can become cost prohibitive for a small clinic or hospital to implement in low- and middle-income countries. In the same vein, training and deploying AI models rely on large annotated datasets and powerful computing resources that may not be universally available. These challenges present a critical access barrier to state-of-the-art imaging technologies and compound healthcare disparities.

We have also been working on making these systems more scalable and economical by creating compact AI models and energy-efficient hardware such as neuromorphic chips. Mafie and Bradley noted that cloud-based platforms can provide a potential pathway for remote access to computational resources, making advanced signal processing analysis in healthcare more accessible to smaller providers. However, guaranteeing fair access to these resources is still a major challenge to overcome.

1.6.3 Future Trends in Biomedical Signal Processing

Emerging trends all have a stake in the future of biomedical signal processing and are likely to improve the capabilities and accessibility of imaging systems. Two focal points include the integration of edge AI and IoT in imaging, and the capability of advanced signal processing to personalize imaging.

1.6.3.1 Image Systems at the Crossroads of Edge AI and IoT

Edge computing and IoT are driving a paradigm shift in biomedical imaging. What is edge AI? Edge AI refers to the deployment of AI algorithms directly on imaging devices (or local servers), which offers reduced latency and eliminates the need to send large volumes of data to centralized servers. This approach is especially useful in situations requiring fast decisions, such as emergency medicine or intraoperative imaging.

With an ideal to create seamless interoperability with other medical devices and electronic health records, imaging systems enabled by the IoT take connectivity to the next level. In this example, wearable US with edge AI can facilitate remote monitoring, continuous US imaging-derived data acquisition, and automatic early complications analysis. These innovations allow the healthcare delivery process to be more efficient, easing the burden on centralized infrastructure.

1.6.3.2 Signal Processing for Personalized Imaging

Despite the recent advances in technology, the true promise of personalized medicine, defined as tailoring healthcare (including prevention and treatment) to the individual characteristics of each patient, is still unfolding in biomedical imaging. In this paradigm, high-end signal processing is key as it enables custom imaging protocols and analysis.

For example, AI-based systems can instantly tailor imaging parameters based on variables specific to the patient, such as the anatomy, age, and medical history of that patient. Techniques like CS and adaptive filtering can enhance an image for a specific diagnostic need and provide clinicians with the ability to get the most pertinent information available for each case. We are more effectively answering diagnostic needs with high precision and low radiation and scan duration, while enhancing patient comfort, using personalized imaging.

The fusion of genomic and imaging data using novel data fusion methodologies is opening additional avenues to precision diagnostics. However, the integration of imaging phenotypes with genetic data will have the potential to offer clinicians actionable insights into disease mechanisms that can facilitate more targeted and effective therapeutic options.

Despite the challenges presented by ethical implications, legal obstacles, and scalability obstacles, the domain of biomedical signal processing can be tremendously innovative and can drive substantial impact.

Edge AI Integrated into IoT and Personalized Imaging Trends on the Horizon in Intelligent Medical Imaging—Intelligent medical imaging has come a long way with many predicting its advancement to new levels if not horizons soon. This advancement has the power to transform healthcare with earlier diagnoses, more precise treatment, and better global health outcomes.

Breakthroughs in signal processing technologies have enabled a rapid advancement of biomedical imaging, transforming the future of medical diagnostics and patient management. In this, we covered several of the main signal processing techniques and their application to certain imaging modalities and have discussed how such techniques have improved image quality and diagnostic accuracy as well as enabled new avenues of temporal and personalized health care.

1.7 Advancements in Signal Processing Techniques and Innovations

To improve the quality, speed, and accuracy of biomedical imaging systems, the most crucial aspect comes from image signal processing. Advanced signal processing methods play a key role at every stage of the imaging pipeline: from data acquisition and preprocessing to real-time processing. Techniques like adaptive filtering for noise reduction have improved the quality of images since they mitigate the effect of artifacts produced by movement or electronic noise. Moreover, image reconstruction algorithms (e.g. CS and WTs) enable shorter scan times to achieve higher fidelity when returning the types of images required by approaches needing long image acquisition times (such as MRI).

Utilizing imaging systems and feature extraction methods (edge detection, texture analysis, etc.) contributed to a significant increase in diagnostic functionality. These methods can detect subtle patterns that would otherwise be missed, thus allowing for the early detection of diseases and the planning of treatment. Real-time processing strategies, such as those leveraging GPU-based parallel computing, have greatly advanced the field of clinical imaging; indeed, such techniques have helped reduce latency and improve imaging system usability in time-sensitive clinical workflows.

Advances in ML and AI have provided an additional layer of transformation, providing automated analysis of images, and improved decision making at the point of diagnosis. For example, DL algorithms have been highly effective in the automation of tasks such as lesion detection, organ segmentation, and disease classification concerning modalities including MRI, CT, and USG. AI algorithms may have the capability to democratize healthcare through decision support systems in resource-constrained environments, increase the diagnostic accuracy of radiographs, and help radiologists and clinicians make more informed decisions. There is a new development, which is the inclusion of quantum signal processing in high-resolution imaging, offering higher precision imaging and furthering the quest for more accurate as well as real-time diagnostics.

Advanced signal fusion and data integration techniques have also been applied to enhance hybrid imaging modalities, such as (PET–CT) and PET–MRI, which integrates complementary imaging technologies and offers integrated information about disease processes. Such advances show the way that cross-modal imaging can give richer data to help guide more accurate clinical decisions.

1.7.1 Future Perspectives on Biomedical Imaging

The future is bright for biomedical imaging, as signal processing technology is an area of continual evolution in a landscape of matching new device technology. The incorporation of edge AI and the IoT into imaging systems is one of the most exciting advancements. With the transition to edge computing, AI algorithms run on imaging devices, revolutionizing healthcare, and offering diagnostic capabilities in real time at the point of care. Minimizing reliance on a centralized cloud not only streamlines decision making but also expedites it, which is especially critical during any crisis, where every second matters. In addition, IoT-based imaging systems will enhance the interaction among devices, resulting in a more interconnected and streamlined medical practice.

Another field where signal processing is about to demonstrate its potential is that of personalized medicine. Individualized assessment of patients' features to develop specific imaging protocols and analysis thus will improve the accuracy of diagnosis, minimizing exposure of the patient to unnecessary doses of radiation and/or time to realize imaging. Moreover, critical for precision diagnostics is the merging of diverse genetic and molecular data into a wide-ranging comprehension of traditional imaging information, potentially leading to paradigm shifts in biomedical imaging systems. Integrating imaging phenotypes and genomic profiles will provide clinicians with enhanced knowledge about disease processes, resulting in more targeted interventions and better patient outcomes.

With the advancement of signal processing, new challenges are expected. It is important to consider the ethical implications of these technologies whether they promote the right outcomes and who gets access to them. Moreover, transparency and interpretability will also continue to be crucial in ensuring trust and reliability in clinical settings as imaging systems become more complex and integrated with AI.

To conclude, the future of biomedical imaging is exceptionally bright. Innovations in signal processing algorithms and techniques inspired by advancements in AI, quantum computing, and multimodal imaging will reshape healthcare delivery, making it more accurate, effective, and readily available. With the maturation of these technologies, healthcare professionals will advance to make early detection and personalized treatment, and ultimately better the lives of patients across the world. This is where the collaboration of scientists, medical professionals, and technology developers will be key to ensure that these developments will be made to their full potential without compromising health or ethics.

1.8 Conclusion

Signal processing innovations have transformed the field of biomedical imaging, revolutionizing healthcare diagnostics, treatment decisions, and research. These innovations have also contributed to the evolution of imaging modalities from X-ray, computiedtomography (CT), magnetic resonance imaging (MRI), and ultrasonography systems to coupling incorporating emerging technologies like AI, quantum computing, and multitudes of data fusion, dramatically increasing image quality, speed, and diagnostic accuracy. All of these advanced signal processing techniques lead to improved noise reduction, image reconstruction, and processing in real time, opening the doors for more patient-specific, accurate, and rapid healthcare interventions. Yet issues around scalability, cost, and ethics continue to be significant. The adoption of edge AI and IoT is suggested as potential solutions to tackle the challenges while addressing the need for healthcare accessibility and equity. By reading this, you are aware that biomedical imaging is rapidly advancing and its place on cutting-edge technology plays a more significant role to really help here and now.

References

1 Akçakaya, M., Yaman, B., Chung, H., and Ye, J.C. (2022). Unsupervised deep learning methods for biological image reconstruction and enhancement: an overview from a signal processing perspective. *EEE Signal Processing Magazine* 39 (2): 28–44. https://doi.org/10.1109/MSP.2021.3119273.

2 Zeng, Y., Dou, T., Ma, L., and Ma, J. (2022). Biomedical photoacoustic imaging for molecular detection and disease diagnosis: 'always-on' and 'turn-on' probes. *Advanced Science* 9 (25): 2202384. https://doi.org/10.1002/advs.202202384.

3 Azad, R., Kazerouni, A., Heidari, M. et al. (2024). Advances in medical image analysis with vision transformers: a comprehensive review. *Medical Image Analysis* 91: 103000. https://doi.org/10.1016/j.media.2023.103000.

4 Bhosale, Y.H. and Patnaik, K.S. (2023). Bio-medical imaging (X-ray, CT, ultrasound, ECG), genome sequences applications of deep neural network and machine learning in diagnosis, detection, classification, and segmentation of COVID-19: a meta-analysis & systematic review. *Multimedia Tools and Applications* 82 (25): 39157–39210. https://doi.org/10.1007/s11042-023-15029-1.

5 Datta, R., Gillette, A., Stefely, M., and Skala, M.C. (2021). Recent innovations in fluorescence lifetime imaging microscopy for biology and medicine. *Journal of Biomedical Optics* 26 (07): 1, 070603–11. https://doi.org/10.1117/1.jbo.26.7.070603.

6 Tanveer, M., Lin, C.T., Ting, C.K., and Andreu-Perez, J. (2023). Guest editorial: special issue on emerging computational intelligence techniques to address challenges in biomedical data and imaging. *IEEE Transactions on Emerging Topics in Computational Intelligence* 7 (2): 292–294. https://doi.org/10.1109/TETCI.2023.3254840.

7 Conze, P.H., Andrade-Miranda, G., Singh, V.K. et al. (2023). Current and emerging trends in medical image segmentation with deep learning. *IEEE Transactions on Radiation and Plasma Medical Sciences* 7 (6): 545–569. https://doi.org/10.1109/TRPMS.2023.3265863.

8 Curtis, A.D., Mertens, A.J., and Cheng, H.L.M. (2023). A predictive signal model for dynamic cardiac magnetic resonance imaging. *Scientific Reports* 13 (1): 1–17. https://doi.org/10.1038/s41598-023-37475-5.

9 Ruiz, J., Schlotthauer, G., Vignolo, L., and Colominas, M.A. (2024). Fully adaptive time-varying wave-shape model: applications in biomedical signal processing. *Signal Processing* 214: 109258. https://doi.org/10.1016/j.sigpro.2023.109258.

10 Melanthota, S.K., Gopal, D., Chakrabarti, S. et al. (2022). Deep learning-based image processing in optical microscopy. *Biophysical Reviews* 14 (2): 463–481. https://doi.org/10.1007/s12551-022-00949-3.

11 Suganyadevi, S., Seethalakshmi, V., and Balasamy, K. (2022). A review on deep learning in medical image analysis. *International Journal of Multimedia Information Retrieval* 11 (1): 19–38. https://doi.org/10.1007/s13735-021-00218-1.

12 Zhou, D., Duan, J., Qin, C., and Luo, G. (2024). Editorial: the combination of data-driven machine learning approaches and prior knowledge for robust medical image processing and analysis. *Frontiers in Medicine* 11: 1434686. https://doi.org/10.3389/fmed.2024.1434686.

13 Warner, E., Lee, J., Hsu, W. et al. (2024). Multimodal machine learning in image-based and clinical biomedicine: survey and prospects. *International Journal of Computer Vision* 132 (9): 3753–3769. https://doi.org/10.1007/s11263-024-02032-8.

14 Ikram, S., Shah, J.A., Zubair, S. et al. (2019). Improved reconstruction of MR scanned images by using a dictionary learning scheme. *Sensors (Basel)* 19 (8): 1–17. https://doi.org/10.3390/s19081918.

15 Selvaganesan, K., Wan, Y., Ha, Y. et al. (2023). Magnetic resonance imaging using a nonuniform Bo (NuBo) field- cycling magnet. *PLoS One* 18 (6 June): 1–21. https://doi.org/10.1371/journal.pone.0287344.

16 Eberhard, M. and Alkadhi, H. (2020). Machine learning and deep neural networks: applications in patient and scan preparation, contrast medium, and radiation dose optimization. *Journal of Thoracic Imaging* 35: S17–S20. https://doi.org/10.1097/RTI.0000000000000482.

17 Sharon, T. and Eldar, Y.C. (2024). Real-time model-based quantitative ultrasound and radar. *IEEE Transactions on Computational Imaging* 10: 1175–1190. https://doi.org/10.1109/TCI.2024.3436537.

18 Huen, I., Zhang, R., Bi, R. et al. (2023). An investigation of signal preprocessing for photoacoustic tomography. *Sensors* 23 (1): 510. https://doi.org/10.3390/s23010510.

19 Han, T., Wu, J., Luo, W. et al. (2022). Review of generative adversarial networks in mono- and cross-modal biomedical image registration. *Frontiers in Neuroinformatics* 16: 933230. https://doi.org/10.3389/fninf.2022.933230.

20 Zhu, X., Huang, Q., DiSpirito, A. et al. (2022). Real-time whole-brain imaging of hemodynamics and oxygenation at micro- vessel resolution with ultrafast

wide-field photoacoustic microscopy. *Light: Science & Applications* 11 (1): 138. https://doi.org/10.1038/s41377-022-00836-2.

21 Chicco, D. and Shiradkar, R. (2023). Ten quick tips for computational analysis of medical images. *PLoS Computational Biology* 19 (1): 1–14. https://doi.org/10.1371/journal.pcbi.1010778.

2

Deep Learning Techniques for Biomedical Imaging

Vandana and Chetna Sharma

Chitkara University Institute of Engineering and Technology, Chitkara University, Punjab, India

2.1 Introduction

Medical science makes extensive use of image and video processing. Both the diagnostic and therapeutic fields heavily rely on biomedical imaging methods. These methods have greatly contributed to the improvement of patient care. With increased precision in illness identification and surgical techniques, image-guided therapy has significantly decreased the high risk of human mistake. Beginning in the 1890s, medical imaging's history grew progressively in the 1980s and has been extensively studied in recent years due to technical developments. This chapter's primary goal is to give beginners an overview of the major biomedical imaging methods and their developments in the modern world [1].

Significant progress has been made in the categorization of medical images over the past few decades utilizing neural networks. A subset of machine learning (ML) is deep learning (DL), which is reliant on examining several layers of perception by creating a hierarchy of traits that differentiate lower and higher levels and can help discover a number of upper-level traits. Image classification, which may be expressed as grouping images into one of several predefined categories, is a major problem in computer vision [2].

Collecting medical images is one of the most challenging duties since both the collection and categorization of health information include security concerns, which is why the specialists' time-consuming details are required. One is to get more information, such as by mass sourcing or by looking at the diagnostic data that is already available in all of its particular modes of resolution. Another approach is to look at various ways to improve the effectiveness of a small dataset, which is crucial when the study experience is applied to a large dataset. Traditional support vector method (SVM) is an ML approach that was employed

Biomedical Imaging Technology: Signal Processing Strategies and Innovations, First Edition.
Edited by Ayush Dogra, Shalli Rani, and Ankita Sharma.

for biomedical image categorization in prior decades [3, 4]. However, SVM has certain drawbacks, including the fact that it takes time to detect and extract visual features, and it performs poorly compared to many other approaches.

The COVID-19 pandemic has significantly impacted healthcare in over half of the countries worldwide, with 53% of cases receiving high blood pressure treatment, 49% diabetes treatment, 42% cancer treatment, 31% cardiovascular crises, and 63% rehabilitation services [5]. The pandemic has also raised concerns for cancer patients, physicians, medical physicists, nurses, and other personnel in providing safety, sanitization, segregation, face/body shielding, maintaining social distance, and setting up radiation infrastructures [6]. Clinical medical physicists are working to reduce worker infection risk in areas such as clinical services, education, informatics, quality assurance, equipment performance evaluation, treatment planning, brachytherapy, in vivo dosimetry, mobility management, and more. Biomedical engineers are saving lives through various means, such as improved nasal swabs, 3D printing, artificial intelligence applications, rapid test kits, face masks, ventilators, oximeters, stronger healthcare infrastructure, imaging modalities, medical equipment to contain the SARS-CoV-2 virus, and vaccine development. However, many illnesses have a high annual death rate, with the top 10 causes accounting for 55% of global fatalities [7–11].

2.2 Overview of DL Architecture in Biomedical Imaging

The number of layers needed to change and transform the data is referred to as "deep" in DL. DL, a subfield of ML that examines many levels of perception by creating a hierarchy of characteristics that differentiate between lower and higher levels, can help identify different upper-level properties. The most sophisticated DL techniques made use of artificial neural networks, particularly CNN, which analyzes pictures using several layers. The goal of each DL layer is to transform the input into a more broad and structured representation. In face recognition, for example, the first layer may encode the pixels and extract the edges; the second layer might then create and interpret the edge layouts; the third layer might then extract the ears and nose; and the fourth layer might decide that a face has now emerged in the image. Since undefined data is more common than recognized and labeled data, DL systems can handle unsupervised learning tasks. This is a significant advantage. Generative Adversarial Networks (GANs) (GANs), autoencoders, Feedforward Neural Networks (FNNs), recurrent neural networks (RNNs), convolutional neural networks (CNNs), encoder–decoder architectures, and deep reinforcement learning (RL) were among the DL designs.

1) A DL design without loop interactions between nodes is the feedforward neural network (FNN) [12]. As the first and most basic kind of DL technology, it was

developed. The hidden layer in this system allows data to go from the input layer to the target layer in a single direction. There are no cycles or loops in the infrastructure. This architecture is for supervised learning.

2) The loop structure of RNNs allows them to store state memory. They are capable of handling patterns in continuous content in an effective and efficient manner [13]. A number of RNN frameworks, including GRUs and LSTMs, have been created to help in sequence property recognition. More precise outputs may be produced by RNNs using their memory to manage the changed sequential inputs. This design is based on supervised learning.
3) CNN is a type of DL that is primarily used for image evaluation. DL is a supervised learning-based architecture that executes tasks like segmentation, image classification, and object recognition that can be used in medical imaging. Its main goal is to enable robots to comprehend real-world scenarios like humans do and to complete tasks using the information gathered [14].
4) The "encoder decoder" architectural design is a more sophisticated analysis that uses the decoding phase to obtain a high-dimensional outcome rather than doing estimation. It is a supervised learning-based architecture. FFNNs, RNNs, and CNNs require dense, recurrent, or convolutional encoders for estimation; these encoders can be combined or adjusted based on the initial dataset used to build a meaningful representation [15].
5) Training networks that provide representative samples are called GANs. Two networks make up the training cycle in its most basic form [16]. To fool the other network, the discriminator, which determines if an image is real or fake, one activity, called a generator, creates new information events. A wide variety of GAN tweaks and variants have been presented in recent years, such as the ability to move image from one region to another and create images from a certain category. The design is built on unsupervised learning.
6) Unsupervised learning is accomplished via Autoencoders, which do not require labeled input. An input layer, which is frequently a hidden layer, and a production layer make up the basic architecture of an unsupervised autoencoder. As the architecture assumes the shape of the bottleneck, autoencoders can be used for dimensional reduction or pretraining. Considering an unseen layered autoencoder, an autoencoder may serve as a feature extraction method and acquire many levels of data representation by incorporating hidden layers [17].
7) Training an entity to maximize reward in a community context is known as deep RL. Neural-network-based training is known as "deep RL." Model-based, policy-based, and value-based are the three categories into which RL architectures fall. Anytime a series of assumptions is anticipated in virtual or real-world contexts, deep reinforcement learning (DLRL) allows us to apply DL [18].

2.3 CNN Architecture

A CNN consists of many hidden layers in addition to inputs and outputs layers. Convolutional levels that relate to transmission or other level components often make up CNN's hidden layers. Multiple convolutions, such as pooling layers, interoperability, and completely connected layers—which are essentially hidden layers—combine with the most popular activation function, RELU [19]. For example, the activation function and final convolution wrap their layer parameters. First, the input is divided into levels and its output is passed on to the following step when convolutional layers are taken into consideration. Only information related to its field of transmission is processed by each convolutional neuron. Applying this framework to medical images is not viable, even if neural networks might be completely integrated with feed-forward to improve capacities and distinguish the data. Due to the extremely large inputs and pixels, where each point is an important component, even with a modest architectural design, a large number of connections will be needed. The convolution process solves the aforementioned issue by lowering the likelihood of empty constraints and allowing for a deeper network with far fewer measurements.

1) **Max Pooling:** Among the several on-linear developments, maximum pooling's debut is arguably the most noticeable. It produces the maximum output by slicing the pixel values in a matrix, dividing the pixel area of an image into smaller parts. Lowering the representation size for feature extraction while preventing overfitting of the customized model is the aim of max-pooling [20].
2) **Activation Function:** The most often used activation function is the RELU, which treats negative values as zero and removes them from the value matrix. It maintains the subfields of convolutional layers while lowering the linear functions of the activation function [21]. Because it trains neural networks more quickly and attains standard accuracy, the RELU function is preferred.
3) **Dropout Layer:** Since most constraints are used by a fully connected layer, a model may be overfit. Dropout is a strategy to prevent overfitting. In the learning phase, independent entities were either retained with probability P or removed from the network with probability (1−P) [22]. In the network design, the dropped entities were reconfigured according to their initial weights. The probability that is most frequently selected for eliminating input node limitations during the learning stages is 0.5. Overfitting can be lessened using dropout if all elements of the training data are removed. Additionally, this method greatly increases the effectiveness of training. Dropout's primary flaw was the inadequate coupling between the nodes and the convolutional layers. Standard probabilistic pooling functions are sometimes replaced by stochastic approaches in stochastic pooling, where the activation among each pooling field is chosen at random using an extended linear allocation determined by activity in the pooling zone [23].

2.4 Basic Concepts in Biomedical Imaging

Integrating DL into biomedical imaging has revolutionized how images are acquired, processed, analyzed, and interpreted. Here are the fundamental principles of biomedical imaging enhanced by DL

2.4.1 Data Representation in Imaging

It is necessary to convert an image to numerical form before processing can begin. The image is divided into tiny components called image elements, or pixels, during this conversion process, which is referred to as digitalization [24]. The most used subdivision technique is the rectangular sample grid. Horizontal lines made up of adjacent pixels are used to divide the picture. Every pixel in the picture has its brightness quantized and sampled. In this step, an integer is created for every pixel, signifying the brightness or darkness of the picture at that particular location. The image is represented as a rectangular array of numbers once every pixel has been analyzed. In addition to an integer value called the gray level, each pixel has a location or address (line or row number, sample or column number) [25]. This digital data gathering may now be processed by a computer. Depending on whether the image's resolution is constant, a digital image is a numerical representation of a two-dimensional image. The term "digital image" usually describes bitmapped or raster images. A digital image is a two-dimensional function $f(x,y)$ that has been optically produced, sampled in a rectangular grid pattern with identical spacing, and quantized at equal gray-level intervals [26]. These days, a digital picture is a rectangular array of quantized sample values in two dimensions. Pixels are the many separate picture components that make up a computer image. A image's quality (resolution) is determined by how many pixels it contains.

2.4.2 Image Reconstruction with DL

AI-powered imagine reconstruction is essential to computer vision. It involves transforming low-resolution, damaged, or incomplete images into their full-resolution, enhanced, or high-resolution equivalents. Image reconstruction techniques are essential to digital forensics, satellite imaging, and medical imaging. It flourishes in fields where results are significantly influenced by visual quality and detail. In the past, picture reconstruction has advanced from basic interpolation techniques. By averaging the values of nearby pixels, missing pixels may be estimated. As a result, more sophisticated methods based on ML and DL were created.

After then, a turning point in the field was the development of CNNs. When it comes to identifying spatial hierarchies in pictures, CNNs excel. Because of this, they may be used to rebuild high-quality pictures from ones with lower resolution. DL models may outperform traditional image resolution methods, as demonstrated by the development of the super-resolution convolutional neural network (SRCNN) [27]. Lastly, by producing new image pixels with unmatched realism, GANs enhanced image reconstruction [28]. New standards for high-resolution outputs have been set by specialized models like Enhanced Super-Resolution Generative Adversarial Networks (ESRGANs) [29].

2.4.2.1 Concept of Image Reconstruction

On the surface, the process of transforming an input image into a better version of itself appears to be rather simple. However, image reconstruction is a complex, multi-step process. Some enhance the image's dimensionality, quality, or detail, for example [30].

The standard procedure is as follows:

1) **Acquisition of Input:** Obtaining a low-resolution, incomplete, or deteriorated input image is the initial step.
2) **Pre-processing:** This comprises reducing image distortions or noise in order to do some initial cleaning. The purpose of this step is to provide the reconstruction algorithm cleaner data.
3) **Transformation:** This stage employs mathematical models to extract characteristics or patterns needed for reconstruction. It also entails transforming the image into a format that facilitates improvements or adjustments.
4) **Reconstruction:** This step enhances and completes the image. Typically, the model based the exact action on the intended result, such as increasing resolution, filling in missing areas, expanding, and so on.
5) **Post-processing:** Involves fine-tuning the reconstructed image to increase visual quality, modify contrast, and sharpen features.

High-resolution imaging is essential for many different applications. In fact, some people utilize it as the overarching objective for all image reconstruction efforts. Additionally, it frequently bases its argument on applications such as urban planning, environmental monitoring, and medical diagnostics. In some cases, a small detail might determine whether a patient receives diagnosis of disease or receives high financing. Similar to this, high-resolution photos can provide information in digital forensics that advances the inquiry. Improving visual material in entertainment may greatly enhance the experience. Traditional imaging methods, such as bicubic interpolation, only use the values of nearby pixels to estimate the values of missing pixels. This method isn't very good at generalizing across a large range of images or at handling more complicated scenarios. Neural networks are used by contemporary DL algorithms to forecast and fill in missing features with significantly greater accuracy. Significantly better generalization results from training with more and more varied datasets. Consequently, this leads to outputs that appear more natural and are less prone to artifacts [31–33].

2.4.3 Image Segmentation

A key method in computer vision and digital image processing is image segmentation. In order to simplify and analyze an image by breaking it down into relevant components, it entails dividing a digital image into several segments (regions or objects). By concentrating on certain regions of interest, this technique increases the efficiency of image processing [34]. The following stages are involved in a typical image segmentation task [35]:

1) Assembles pixels in an image according to common traits like texture, color, or intensity.
2) Gives each pixel a label denoting its association with a particular section or item.
3) A segmented image is the end product, which is frequently shown as a mask or overlay emphasizing the various parts.

2.4.3.1 Traditional Image Segmentation Techniques

The following techniques are used in traditional image segmentation methods, which served as the basis for contemporary DL algorithms. These techniques rely more on mathematical operations, image processing principles, and heuristics to divide an image into meaningful regions.

1) **Thresholding:** This technique divides image pixels into foreground and background groups according to intensity levels after choosing a threshold value.
2) **Edge Detection:** This technique detects sudden discontinuities or changes in image intensity. It makes use of algorithms such as Laplacian edge detectors, Canny, and Sobel.
3) **Region-Based Segmentation:** To deal with noise and irregularities in the image, this technique divides the image into smaller sections and repeatedly combines them according to predetermined color, intensity, and texture parameters.
4) **Clustering Algorithm:** This technique groups object pixels in an image into clusters according to comparable characteristics like color or texture using algorithms like *K*-means or Gaussian models.
5) **Watershed Segmentation:** Watershed lines are identified using pixel intensity and connectivity, similar to water flowing down distinct valleys, in a manner similar to that of a topographical map. Although they have limitations, these conventional approaches give fundamental image segmentation techniques that serve as a basis for more sophisticated approaches.

2.4.3.2 DL Image Segmentation Models

A potent method for automatically dividing an image into several segments and extracting characteristics for precise analysis and segmentation tasks is DL image segmentation models, which make use of the neural network architecture. Some of the well-liked DL models for image segmentation are listed below:

1) **U-Net:** This model effectively segments medical images using a U-Shaped network. The approach offers accurate segmentation and is highly effective when working with tiny amounts of data [36].
2) **Fully Convolutional Network (FCN):** This model can produce spatial maps and interpret images of any size. Convolutional layers are used in place of completely linked layers in a traditional CNN to accomplish this. This facilitates pixel-by-pixel segmentation of an entire image [36].
3) **SegNet:** This model has an encoder–decoder network for tasks like object identification and scene comprehension. Here, the encoder records the image's context, and the decoder uses that context to precisely locate and segment the objects [37].
4) **DeepLab:** Using atrous convolutions to capture multi-scale context with many simultaneous filters is DeepLab's primary characteristic [38].
5) **Mask R-CNN:** By including a branch for segmentation mask prediction in addition to bounding box regression, this model expands upon the faster R-CNN object identification framework [39].
6) **Vision Transformer (ViT):** A novel approach to image segmentation that uses transformers. To comprehend the image's global context, the image is separated into patches and processed one after the other [40].

2.4.4 Image Registration

In medical imaging, image registration is a crucial issue. It offers a wide range of possible uses in clinical diagnostics, including the diagnosis of illnesses of the heart, retina, pelvis, kidneys, abdomen, liver, and tissue [41]. In order to track even the smallest variations between two pictures, they must be aligned into a single coordinate system. Remote sensing (multispectral categorization), environmental monitoring, change detection, picture mosaicing, weather forecasting, creating super-resolution images, and incorporating data into geographic information systems (GIS) are some of the uses for image registration. In order to track tumor progression, gather more thorough patient data, and confirm therapy, image registration is used in medicine to merge data from many modalities, including computer tomography (CT) and magnetic resonance imaging (MRI). The steps involved in image registration are as follows [42].

1) **Feature Detection:** In both reference and felt images, salient and distinguishing elements are found.

2) **Feature Matching:** The established relationship between the reference and sensed image's characteristics.
3) **Determining the Kind and Parameters** of the so-called mapping functions that correspond the reference image to the sensed image is a necessary step in estimating the transform model.
4) **Image Modification and Re-sampling:** The identified image is altered using the mapping functions.

2.4.5 Diagnosis and Classification

The vital function that image categorization plays in contemporary technologies makes it stand out. It entails using previously collected training data from previously labeled images to provide a label or tag to an entire image. To get the best label for the entire image, the procedure involves pixel-level image analysis, even if it can seem straightforward at first [43]. This gives us useful information and insights that help us make wise decisions and achieve practical results. To prevent inconsistencies in the data, we must ensure that data labeling is done correctly throughout the training stage. Publicly accessible datasets are frequently employed in the model training stage to enable precise data classification.

2.4.5.1 Types of Image Classification

There are several kinds of image categorization techniques that may be used, depending on the issue. These include hierarchical, multiclass, multilabel, and binary [44].

1) **Binary:** Binary classification divides unknown data points into two groups and labels images using an either-or logic. Binary classification is used to address various issues that need yes/no responses, such as classifying benign or malignant tumors and analyzing product quality to determine whether or not it contains faults.
2) **Multiclass:** As the name implies, multiclass groups things into three or more classes, whereas binary classification is used to differentiate between two classes of objects. It is highly helpful in a variety of fields, such as medical diagnosis (categorizing disorders), natural language processing (NLP), and sentiment analysis (when more than two emotions are present).
3) **Multilabel:** Multilabel classification enables the object to be allocated to numerous labels, in contrast to multiclass classification, which assigns each image to precisely one class. For instance, you might have to categorize images with many hues. Depending on how creative you are with fruit salads, an image of one will feature red, orange, yellow, purple, and other hues. As a result, different colors will be used to designate the same image.

4) **Hierarchical:** The process of classifying classes into a hierarchical structure according to their commonalities is known as hierarchical classification; a lower-level class is more detailed and concrete, while a higher-level class indicates larger categories. Let's return to our fruits and use a delicious example to help us comprehend the idea.

2.4.5.2 Working of Image Classification

In order to make an informed prediction, computer vision must first analyze the image's different components before determining and classifying the image as a whole. For this reason, methods for classifying images use pixels to evaluate them, viewing them as a collection of matrices whose size depends on the image resolution. Pixels from the digital image are extracted and categorized into "classes." The approach will then differ based on the algorithm, but let's take a more general look at how ML functions before we look at the different algorithms. To ensure that the image is not entirely left up to the final classifier, instead, the selected method will transform it into a number of significant characteristics. These characteristics help the classifier identify the image's subject and the class to which it belongs. In general, the method of classifying images looks like this [45] (Figure 2.1).

2.4.5.2.1 Image Pre-processing The data preparation step involves putting techniques into place to enhance the quality of your image data and get it ready for the next steps. The following are some instances of possible inclusions in the image processing step [46]:

1) **Image Resizing:** This involves altering the width and height of the image. In order to simplify the calculation of images for further processing, they are frequently scaled to a common size [47].
2) **Image Cropping:** When an image has extraneous or irrelevant elements that might impair the model's performance (such the backdrop or borders), it is preferable to crop the image so that only the required portions remain [48].
3) **Image Normalization:** This technique is used to bring the values of an image's pixels into line with a standard distribution. To make the image attributes (such as brightness or contrast) more appropriate for analysis, it involves rescaling pixel values to a specific range, removing the mean and dividing it by the standard deviation of pixel values, or employing histogram equalization algorithms [49].

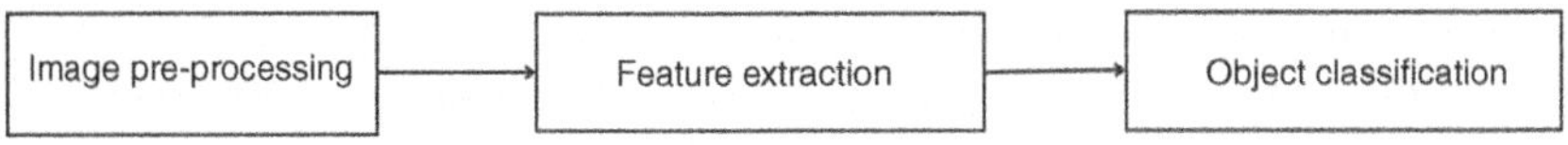

Figure 2.1 Steps involves in image classification.

4) **Noise Reduction:** The accuracy and performance of models can be impacted by the noise present in digital image. To increase image quality, this calls for image filtering methods like Weiner, median, or Gaussian filtering. By blurring or smoothing the image while maintaining edges and features, these methods lower noise [50].
5) **Data Augmentation:** This is the process of generating additional image variants through image manipulations including flipping, zooming, rotating, and brightness and contrast adjustments. In order to improve the accuracy and resilience of your image classification model, data augmentation aims to expand and diversity the training dataset [51].

2.4.5.2.2 Feature Extraction A crucial stage in image classification is feature extraction, which finds visual patterns within an image that might be utilized to differentiate one object from another. There is a distinct class distinction since the patterns are frequently exclusive to the class of photographs in question. The computer may utilize these important image characteristics to classify new images it has never seen before after it has learned and identified them in the training data [52]. When classifying photos of dogs and cats, specific patterns, such as body size and shape, nose/eye form and color, ear position and shape, and hair texture and color, can be used as characteristics to differentiate between the two groups. A key component of image analysis is model training, which is the process of extracting features from the dataset. By concentrating on the most significant and pertinent elements of the data, feature extraction improves the performance of ML models. Without it, models would have to examine whole images, which would take a very long time and a lot of computer power. Techniques for feature extraction include texture analysis, edge detection, and DL algorithms like CNN.

Finding the borders between areas in an image is known as edge detection, and it is used to learn more about the structure and form of things. There are a number of edge detection approaches, including gradient operators, derivation, and some more sophisticated ones. The process of identifying recurring patterns in an image is known as texture analysis, and it may be used to determine whether texture is present and to differentiate between various item surfaces or materials. Identifying cancers in medical imaging is a well-known practical use of texture analysis; clinicians may diagnose the kind of tumor by comparing the texture of diseased and healthy tissue. CNNs, which are frequently employed to enable algorithms to learn directly from raw data, are another noteworthy DL approach for feature extraction. Here, the network identifies the most significant patterns for several picture classes after learning on a sizable dataset of annotated images [53].

2.4.5.2.3 Image Classification and Object Detection Image categorization, object recognition, and object localization may all be jumbled in your thoughts, which is quite normal if you are unfamiliar with these ideas. In actuality, they are critical components of computer vision and image annotation, with their own set of peculiarities [54].

The process of giving a whole image a particular label is known as image categorization. However, object localization focuses on accurately locating and identifying the primary item or regions of interest in an image, going beyond categorization. Object localization gives precise spatial information by enclosing these items in bounding boxes, enabling more focused analysis. Contrary to image classification, which classifies the whole image, object detection locates objects inside an image and labels them with the appropriate label.

Object detection identifies the target objects inside an image, labels them, and indicates their location, as the name suggests. One of the most often used tools for item recognition is the "bounding box," which shows the location of a certain object on an image together with its name. In essence, object detection blends object location with image categorization [55].

2.4.5.2.4 Algorithms and Models The two most prominent types of image classification—supervised and unsupervised classification—will be examined because there isn't a single, simple method. Image categorization is a particularly noteworthy topic in computer vision due to its indispensable function in contemporary technologies. It entails giving a label or tag to a whole image using training data from previously labeled images. The approach involves pixel-level image analysis to identify the best label for the entire image, even if it can seem straightforward at first. The useful information and insights as a result helps us make wise judgments and achieve practical results [56]. However, in order to prevent data inconsistencies, ensure that data labeling is done correctly throughout the training phase. In the model training stage, publicly accessible datasets are frequently employed to enable precise data tagging.

2.4.6 Functional and Molecular Imaging

Molecular imaging, a non-invasive technique for analyzing cellular and subcellular processes, is revolutionizing therapeutic and scientific fields. It involves advancements in engineering, molecular biology, chemistry, immunology, and genetics. These innovations aim to develop non-invasive strategies for easy identification, risk stratification, and monitoring therapy response with high sensitivity and specificity. The combination of imaging and therapy is expected to expand as molecular therapeutics and imaging of molecular and cellular activities merge. Recent advancements include magnetic resonance, optical, CT, ultrasonographic, PET, PET coupled with CT, and single-photon emission computed tomography [57, 58].

2.4.7 Explainability and Interpretability

The capacity to comprehend how an AI model makes decisions is known as interpretability. An interpretable model gives information about the connections between inputs and results and operates transparently. A human may provide a clear and understandable explanation of an interpretable algorithm. For humans to comprehend and have faith in artificial intelligence models, interpretability is crucial [59]. The capacity to communicate an AI model's decision-making process in a way that the user can comprehend is known as explainability. Users may better understand why a model generated a certain outcome by using an explainable model, which offers a concise and understandable explanation of the decisions taken. Stated differently, explainability is concerned with the justification and reasoning behind the decisions made by an algorithm [60]. Although both interpretability and explainability are essential for understanding AI models, there are a number of important differences between the two [61]:

1) **Level of Detail:** Explainability focuses on elucidating the choices taken, whereas interpretability focuses on comprehending the inner workings of the models. As such, interpretability needs more information than explainability.
2) **Model Complexity:** Due to their sophisticated structure and the relationships between many components, more complex AI models, such deep neural networks, can be challenging to understand. Explainability, which focuses on elucidating judgments rather than comprehending the model itself, could be more practical in certain situations.
3) **Communication:** Explainability is more concerned with informing end users of model decisions, whereas interpretability is more concerned with how AI specialists and researchers comprehend the model. Explainability hence necessitates a more straightforward and user-friendly information presentation.

2.4.7.1 Significance of Interpretability and Explainability

This is particularly crucial in professions like medicine where decisions can directly affect people's lives. Therefore, knowing how ML algorithms operate may assist to guarantee that the choices these algorithms make are correct and that mistakes are kept to a minimum. For AI models to be dependable, safe, and compliant with context-appropriate ethical standards, both interpretability and explainability are crucial. These ideas are significant for the following reasons [61, 62]:

1) **Responsibilities:** Users can comprehend the decision-making process and consider the effects of the model's choices when it is interpretable and explicable. This is essential to guaranteeing openness and accountability in the application of AI.
2) **Trust:** Users' confidence in the choices made by AI-based systems may be bolstered by comprehending AI models through interpretability and

explainability. Users are more inclined to believe a model's suggestions when they comprehend how it operates and the reasoning behind its choices.

3) **Adaptation:** Developers may have a better understanding of model performance and pinpoint any issues or areas in need of development by using an interpretable and explainable model. This makes it easier for AI models to be adjusted and improved over time.
4) **Regulatory Compliance:** More openness in the AI models' decision-making process is frequently necessary to comply with data protection and AI ethics laws. For models to meet these standards, interpretability and explainability are crucial.
5) **Bias Reduction:** Bias in data and decision-making may be identified and reduced by comprehending how AI models function through interpretability and explainability. By doing this, it will be possible to guarantee that AI models are more egalitarian and do not discriminate based on sensitive traits like gender, race, or handicap [63].

2.5 Future Study and Application of Image Processing in Biomedical

Medical physics, which includes therapeutic radiological physics, medical, nuclear, and medical health physics, is the application of physics to medicine [64]. The need for comprehensive imaging facilities and precise explanations is a basic aspect of medical physics [64]. The discovery of X-rays, also referred to as medicine in radiation, marked the beginning of medical physics and imaging [65]. More than a century ago, radiation therapy (RT) was initially applied to treat cancer. Significant advancements have since been achieved to increase this modality's efficacy and reduce its adverse effects [66]. RT is a type of radiation medicine used to treat a range of cancer patients. It includes brachytherapy and external beam radiation treatment.

High doses of radiation are used in RT, commonly known as radiotherapy, to destroy cancer cells and reduce tumor size [67]. Throughout the history of radiation treatment, a variety of equipment have been employed to generate radiation beams [68]. External beam radiation therapy (EBRT) uses high-energy electron or X-ray beams to treat cancer [69]. A procedure called brachytherapy involves implanting radioactive material into the patient's body. Brachytherapy, often known as internal RT, is one kind of radiation treatment used to treat cancer [67]. In order to improve the outcome of cancer treatment, certain special techniques are used, such as "stereotactic irradiation, total body irradiation, total skin electron irradiation, intraoperative radiotherapy, endocavitary rectal irradiation, conformal radiotherapy, image-guided radiotherapy, adaptive radiotherapy, respiratory-gated radiotherapy, and PET/CT scanners PET/CT image fusion" [70].

2.6 Conclusion

Biomedical image processing continues to pave the way for groundbreaking innovations in medical science, offering valuable insights and enhancing clinical practices. The integration of computer-aided diagnostic processing has revolutionized clinical routines, streamlining the diagnostic process and contributing to improved patient outcomes. As technology continues to evolve, addressing the challenges related to processing and analyzing vast volumes of medical images remains a top priority for the medical imaging community. With continued research and development, medical image processing will continue to drive advancements in the medical field and shape the future of healthcare. We have given an outline of several fundamental ideas in medical image processing in this chapter. It is crucial to emphasize that none of these issues have been fully solved and that the algorithms we discussed still have a lot of potential for development. Particularly, segmentation is still a difficult operation that frequently requires interactive human involvement to produce the best results. Despite these difficulties, significant strides in autonomous medical image analysis have been made lately thanks to breakthroughs in hardware, acquisition strategies, signal processing methodologies, and mathematical frameworks. Particularly for diverse image processing tasks, curvature-driven flows have demonstrated to be highly effective, having a substantial impact on the technological underpinning. Medical imaging poses substantial mathematical challenges, cutting across various branches of mathematics. As we continue to progress in this field, every bit of help and innovation is valuable to push the boundaries and improve the state of the art. In conclusion, while there is still much to be accomplished, advancements in medical image processing have been made possible through interdisciplinary efforts, and ongoing research promises further enhancements in the future.

References

1 Paul, S., Saikia, A., Majhi, V., and Pandey, V.K. (2022). *Introduction to Biomedical Instrumentation and Its Applications*. Academic Press.

2 Abdou, M.A. (2022). Literature review: efficient deep neural networks techniques for medical image analysis. *Neural Computing and Applications* 34 (8): 5791–5812.

3 Kurani, A., Doshi, P., Vakharia, A., and Shah, M. (2023). A comprehensive comparative study of artificial neural network (ANN) and support vector machines (SVM) on stock forecasting. *Annals of Data Science* 10 (1): 183–208.

4 Huish, R. (2024). *Come from Away: How Rolling Stigma from COVID-19 Shaped Global Health for the 21st Century*. Cham: Springer Nature.

5 Hegelund, M.H., Fjordside, L., Faurholt-Jepsen, D. et al. (2023). Opportunistic non-communicable diseases in times of COVID-19. *APMIS* 131 (1): 13–18.

6 Aden, D., Zaheer, S., and Raj, S. (2022). Challenges faced in the cancer diagnosis and management—COVID-19 pandemic and beyond—lessons for future. *Heliyon* 8 (12): e12091.

7 Coleman, C.N., Buchsbaum, J.C., Prasanna, P.G. et al. (2021). Moving forward in the next decade: radiation oncology sciences for patient-centered cancer care. *JNCI Cancer Spectrum* 5 (4): pkab046.

8 Iijima, K., Nakayama, H., Nakamura, S. et al. (2024). Analysis of human errors in the operation of various treatment planning systems over a 10-year period. *Journal of Radiation Research* 65 (5): 603–618.

9 Guvener, O., Eyidogan, A., Oto, C., and Huri, P.Y. (2021). Novel additive manufacturing applications for communicable disease prevention and control: focus on recent COVID-19 pandemic. *Emergent Materials* 4: 351–361.

10 Ogidi, O.I. The utilization of biomaterials in diagnostic, preventive, and therapeutic approaches for the management of COVID-19. In: *Advances in Sustainable Biomaterials* (ed. D.K. Rajak, A. Kumar, and A. Kumar), 209–232. Boca Raton: CRC Press.

11 Abere, O.J. (2023). Mortality differentials in gender, income level and cause of death. *ANVESHAK-International Journal of Management* 12 (1): 44–72.

12 Kamali, K. (2024). Deep learning (Part 1)-Feedforward neural networks (FNN) (Galaxy Training Materials). https://training.galaxyproject.org/training-material/topics/statistics/tutorials/FNN/tutorial.html (accessed 12 September 2025).

13 Kasongo, S.M. (2023). A deep learning technique for intrusion detection system using a recurrent neural networks based framework. *Computer Communications* 199: 113–125.

14 Purwono, P., Ma'arif, A., Rahmaniar, W. et al. (2022). Understanding of convolutional neural network (CNN): a review. *International Journal of Robotics and Control Systems* 2 (4): 739–748.

15 Jadouli, A. and Amrani, C.E. (2024). Parallel multi-path feed forward neural networks (PMFFNN) for long columnar datasets: a novel approach to complexity reduction. arXiv preprint arXiv:2411.06020.

16 Alqahtani, H., Kavakli-Thorne, M., and Kumar, G. (2021). Applications of generative adversarial networks (GANS): an updated review. *Archives of Computational Methods in Engineering* 28: 525–552.

17 Berahmand, K., Daneshfar, F., Salehi, E.S. et al. (2024). Autoencoders and their applications in machine learning: a survey. *Artificial Intelligence Review* 57 (2): 28.

18 Tang, C., Abbatematteo, B., Hu, J. et al. (2024). Deep reinforcement learning for robotics: a survey of real-world successes. *Annual Review of Control, Robotics, and Autonomous Systems* 39 (27): 28694–28698.

19 Krichen, M. (2023). Convolutional neural networks: a survey. *Computers* 12 (8): 151.

20 Zafar, A., Aamir, M., Mohd Nawi, N. et al. (2022). A comparison of pooling methods for convolutional neural networks. *Applied Sciences* 12 (17): 8643.

21 Yuen, B., Hoang, M.T., Dong, X., and Lu, T. (2021). Universal activation function for machine learning. *Scientific Reports* 11 (1): 18757.

22 Pham, H. and Le, Q. (2021, May). Autodropout: learning dropout patterns to regularize deep networks. *Proceedings of the AAAI Conference on Artificial Intelligence* (Vol. 35, No. 11, pp. 9351–9359).

23 Wang, S.H., Khan, M.A., and Zhang, Y.D. (2021). VISPNN: VGG-inspired stochastic pooling neural network. *Computers, Materials & Continua* 70 (2): 3081–3097.

24 Vrana, J. and Singh, R. (2021). Digitization, digitalization, and digital transformation. In: *Handbook of Nondestructive Evaluation 4.0* (ed. N. Meyendorf, N. Ida, R.(R.) Singh, and J. Vrana), 1–17. Springer Nature.

25 Iqbal, N. and Hanif, M. (2021). An efficient grayscale image encryption scheme based on variable length row-column swapping operations. *Multimedia Tools and Applications* 80 (30): 36305–36339.

26 Castleman, K.R. (2023). Image digitization and display. In: *Microscope Image Processing* (ed. F. Merchant and K. Castleman), 27–45. Academic Press.

27 Zheng, Y. (2023). Hexagonal diffractive optical elements and phase retrieval algorithms. Doctoral dissertation, King Abdullah University of Science and Technology.

28 Dash, A., Ye, J., and Wang, G. (2023). A review of generative adversarial networks (GANs) and its applications in a wide variety of disciplines: from medical to remote sensing. *IEEE Access* 12: 18330–18357.

29 Song, J., Yi, H., Xu, W. et al. (2023). ESRGAN-DP: enhanced super-resolution generative adversarial network with adaptive dual perceptual loss. *Heliyon* 9 (4): e15134.

30 Lepcha, D.C., Goyal, B., Dogra, A., and Goyal, V. (2023). Image super-resolution: a comprehensive review, recent trends, challenges and applications. *Information Fusion* 91: 230–260.

31 Yaqub, M., Jinchao, F., Arshid, K. et al. (2022). Deep learning-based image reconstruction for different medical imaging modalities. *Computational and Mathematical Methods in Medicine* 2022 (1): 8750648.

32 Zhu, R.J., Zhao, Q., Li, G., and Eshraghian, J.K. (2023). SpikeGPT: generative pre-trained language model with spiking neural networks. arXiv preprint arXiv:2302.13939.

33 Zangana, H.M., Mohammed, A.K., Sallow, Z.B., and Mustafa, F.M. (2024). Exploring image representation and color spaces in computer vision: a comprehensive review. *The Indonesian Journal of Computer Science* 13 (3): 4261–4272.

34 Wang, R., Lei, T., Cui, R. et al. (2022). Medical image segmentation using deep learning: a survey. *IET Image Processing* 16 (5): 1243–1267.

35 Azad, R., Aghdam, E.K., Rauland, A. et al. (2024). Medical image segmentation review: the success of u-net. *IEEE Transactions on Pattern Analysis and Machine Intelligence* 46: 10076–10095.

36 Wu, J., Liu, B., Zhang, H. et al. (2021). Fault detection based on fully convolutional networks (FCN). *Journal of Marine Science and Engineering* 9 (3): 259.

37 Bansal, T., Gupta, S., and Jindal, N. (2023, December). Segmentation techniques for detection of tuberculosis using deep learning: a review. In: *2023 3rd International Conference on Smart Generation Computing, Communication and Networking (SMART GENCON)*, 1–6. IEEE.

38 Wang, J. and Liu, X. (2021). Medical image recognition and segmentation of pathological slices of gastric cancer based on Deeplab v3+ neural network. *Computer Methods and Programs in Biomedicine* 207: 106210.

39 Kumar, D. and Kukreja, V. (2024). A novel cross-SVM framework for wheat rust disease recognition through multimodal fusion of images, mask R-CNN, and DenseNet.

40 Yin, H., Vahdat, A., Alvarez, J.M., et al. (2022). A-ViT: adaptive tokens for efficient vision transformer. In *Proceedings of the IEEE/CVF Conference on Computer Vision and Pattern Recognition* (pp. 10809–10818).

41 Sun, H. (2023). A review of 3D-2D registration methods and applications based on medical images. *Highlights in Science, Engineering and Technology* 35: 200–224.

42 Suganyadevi, S., Seethalakshmi, V., and Balasamy, K. (2022). A review on deep learning in medical image analysis. *International Journal of Multimedia Information Retrieval* 11 (1): 19–38.

43 Mittal, S., Srivastava, S., and Jayanth, J.P. (2023). A survey of deep learning techniques for underwater image classification. *IEEE Transactions on Neural Networks and Learning Systems* 34 (10): 6968–6982.

44 Salvi, M., Acharya, U.R., Molinari, F., and Meiburger, K.M. (2021). The impact of pre-and post-image processing techniques on deep learning frameworks: a comprehensive review for digital pathology image analysis. *Computers in Biology and Medicine* 128: 104129.

45 Park, K., Chae, M., and Cho, J.H. (2021). Image pre-processing method of machine learning for edge detection with image signal processor enhancement. *Micromachines* 12 (1): 73.

46 Giełczyk, A., Marciniak, A., Tarczewska, M., and Lutowski, Z. (2022). Pre-processing methods in chest X-ray image classification. *PLoS One* 17 (4): e0265949.

47 Billa, N.R., Das, B.P., Biswal, M., and Okade, M. (2024). CNN based image resizing forensics for double compressed JPEG images. *Journal of Information Security and Applications* 81: 103693.

48 Zhan, W., Luo, F., Luo, H. et al. (2024). Time-series-based spatiotemporal fusion network for improving crop type mapping. *Remote Sensing* 16 (2): 235.

49 Vu, V.H. (2024). Content-based image retrieval with fuzzy clustering for feature vector normalization. *Multimedia Tools and Applications* 83 (2): 4309–4329.

50 Inoue, Y., Itoh, H., Hata, H. et al. (2024). Noise reduction in brain CT: a comparative study of deep learning and hybrid iterative reconstruction using multiple parameters. *Tomography* 10 (12): 2073–2086.

51 Tian, J., Jiang, Y., Zhang, J. et al. (2024). A novel data augmentation approach to fault diagnosis with class-imbalance problem. *Reliability Engineering & System Safety* 243: 109832.

52 Talukder, M.A., Islam, M.M., Uddin, M.A. et al. (2024). Machine learning-based network intrusion detection for big and imbalanced data using oversampling, stacking feature embedding and feature extraction. *Journal of Big Data* 11 (1): 33.

53 Kaushik, P. and Sharma, P. (2024, November). Cutting-edge convolutional neural network autoencoders for anomaly detection in ECG signals: advancements in early cardiac diagnosis. In *2024 International Conference on Cybernation and Computation (CYBERCOM)* (pp. 340–344). IEEE.

54 Wang, J., Hu, F., Abbas, G. et al. (2024). Enhancing image categorization with the quantized object recognition model in surveillance systems. *Expert Systems with Applications* 238: 122240.

55 Kang, W., Liu, G., Shah, M., and Yan, Y. (2025). SegVG: transferring object bounding box to segmentation for visual grounding. In: *European Conference on Computer Vision*, 57–75. Cham: Springer.

56 Zhao, M., Hu, X., Zhang, L. et al. (2024). Beyond pixel-level annotation: exploring self-supervised learning for change detection with image-level supervision. *IEEE Transactions on Geoscience and Remote Sensing* 62: 1–16.

57 Van Hoogstraten, S.W.G., Kuik, C., Arts, J.J.C., and Cillero-Pastor, B. (2024). Molecular imaging of bacterial biofilms—a systematic review. *Critical Reviews in Microbiology* 50 (6): 971–992.

58 Puccetti, M., Pariano, M., Schoubben, A. et al. (2024). Biologics, theranostics, and personalized medicine in drug delivery systems. *Pharmacological Research* 201: 107086.

59 Xua, B. and Yang, G. (2024). Interpretability research of deep learning: a literature survey. *Information Fusion* 115: 102721.

60 Chander, B., John, C., Warrier, L., and Gopalakrishnan, K. (2024). Toward trustworthy artificial intelligence (TAI) in the context of explainability and robustness. *ACM Computing Surveys* 57 (6): 1–49.

61 Şahin, E., Arslan, N.N., and Özdemir, D. (2024). Unlocking the black box: an in-depth review on interpretability, explainability, and reliability in deep learning. *Neural Computing and Applications* 37 (2): 859–965.

62 Hosain, M.T., Jim, J.R., Mridha, M.F., and Kabir, M.M. (2024). Explainable AI approaches in deep learning: advancements, applications and challenges. *Computers and Electrical Engineering* 117: 109246.

63 Tran, T.T. (2024). Explainability vs. Interpretability and methods for models' improvement. Doctoral dissertation, Hochschule für Angewandte Wissenschaften Hamburg.

64 Gambo, N. and Shehu, M. (2024). The role of diagnostic medical physics in medicine: an overview. *Sahel Journal of Life Sciences FUDMA* 2 (1): 103–109.

65 Kitson, S.L. (2024). Modern medical imaging and radiation therapy. Cyber Security| Big Data| AI. Open Medicine Science.

66 Abbas, N.A.H.R.J., Nasser, A.M.J., Hussein, S.S.R., and Muheisen, H.M.A. (2024). Functional nanomaterials for X-ray triggered cancer therapy: chemotherapy and brachytherapy application techniques. *Current Clinical and Medical Education* 2 (10): 72–91.

67 Abd, F.S.H., Kadhum, Z.A., Abd-Zaid, F.A.Z.H. et al. (2024). Radiation therapy of cancers, X-ray beams, therapeutic photon beams, and brachytherapy application techniques. *Journal of Current Medical Research and Opinion* 7 (6).

68 Xiao, Q. and Li, G. (2024). Application and challenges of statistical process control in radiation therapy quality assurance. *International Journal of Radiation Oncology* Biology* Physics* 118 (1): 295–305.

69 Safavi-Naeini, M., Boyle, T.P., Sheehy, S., and Liyanage, S. (2024). Big science medical applications from accelerator physics. In: *Big Science, Innovation, and Societal Contributions: The Organisations and Collaborations in Big Science Experiments*, vol. 220 (ed. M. Nordberg, S. Liyanage, and M. Streit-Bianchi). Oxford University Press.

70 Grosu-Bularda, A., Lita, F.F., Hodea, F.V. et al. (2024). Navigating the complexities of radiation injuries: therapeutic principles and reconstructive strategies. *Journal of Personalized Medicine* 14 (11): 1100.

3

Advanced Methods and Approaches in Image Reconstruction

Navneet Kaur[1] *and Gurbinder Singh Brar*[2]

[1] *School of Computer Application, Lovely Professional University, Jalandhar, Punjab, India*
[2] *School of Computer Science & Engineering, Lovely Professional University, Jalandhar, Punjab, India*

3.1 Introduction

From early uses of X-rays for fracture diagnosis and foreign body detection, biomedical imaging has grown into a collection of effective methods for patient care, biological structure, and function research, and answering basic biomedical questions. A variety of non-invasive techniques for examining entire three-dimensional bodies have been made possible by advances in technology in digital radiography, X-ray computed tomography (CT), nuclear imaging (including positron emission tomography, or PET), ultrasound, and optical and magnetic resonance imaging (MRI). Biomedical imaging science has developed into a unique and cohesive body of knowledge in recent years, and it now plays a key role in a large portion of medical research. As imaging technology and allied sciences like molecular biology and nanotechnology continue to advance, imaging's uses have been extended to new fields like the study of gene expression or the brain's functional organization. The justification for the growing involvement of imaging specialists in biomedical research and clinical practice is made clear in this volume, which also describes several significant recent advances in biomedical imaging technology [1]. The evolution of imaging from qualitative graphical representations of anatomy to a field that provides quantitative measurements of many biomedical processes is specifically illustrated by the many examples provided. Whatever the name of the specific modality, imaging science refers to a dynamic, developing field of multi- and interdisciplinary activities that encompass at least three different forms of endeavor. Imaging scientists are always creating new equipment or adjusting methods to produce, for instance, images with higher spatial resolution, acquired in shorter amounts of time, or that

Biomedical Imaging Technology: Signal Processing Strategies and Innovations, First Edition.
Edited by Ayush Dogra, Shalli Rani, and Ankita Sharma.

utilize some new kind of tissue interaction with an externally applied energy field. The first goal of imaging science is to develop and improve imaging technology to provide new or better information (e.g., new ways like functional MRI) to detect and map neural activity in the brain [2].

In recent years, imaging science has become much more significant and expansive. All modalities continue to see significant technological advancements, and the creation of faster, more potent computers has produced sophisticated image analysis techniques and processing algorithms that can be used to extract useful, quantitative information from images, such as measurements of the sizes of particular brain structures. Informatics techniques for mining big image collections are increasingly demonstrating the relationship between multidimensional data, including structure and genotype. Image reconstruction in biomedical imaging continues to present difficulties despite significant advancements. Image quality can be compromised by noise and artifacts produced during data gathering, which can mask important details. The identification of tiny details may be hampered by limited resolution, particularly in modalities like PET. Another major problem is computational efficiency since many reconstruction techniques demand a lot of computing power, which can be problematic in contexts with limited resources or for applications that need quick results [3].

Applications for real-time imaging, including point-of-care diagnostics or interventional procedures, provide another level of complexity by requiring accurate and quick picture reconstruction. To overcome these obstacles and drive innovation in picture reconstruction, signal processing is essential. Modern signal processing methods make it easier to recover useful information from noisy or insufficient datasets, improving the reconstructed images' readability and clarity. Recent developments have greatly increased picture resolution, decreased noise, and sped up processing efficiency. These breakthroughs include compressed sensing (CS), model-based techniques, and machine-learning-driven methodologies. Combining the advantages of both classical and contemporary approaches, hybrid approaches that combine data-driven algorithms with physics-based modeling are becoming increasingly potent instruments. Biomedical imaging is becoming more capable thanks to the combination of signal processing and emerging technologies like edge computing, quantum computing, and artificial intelligence (AI). These developments have the potential to overcome current constraints and pave the way for the creation of more effective, accurate, and widely available next-generation imaging systems. Beyond technical developments, image reconstruction has significant effects on patient care since better diagnosis and treatment results are directly correlated with higher-quality imaging. Ongoing innovation in this area improves healthcare delivery, develops medical science, and ultimately saves lives [4].

3.1.1 Fundamental Principles of Image Reconstruction

Image reconstruction is the process of turning a difficult-to-understand set of data into an easier-to-understand target image. In general, there are two categories of image reconstruction problems. In the first type, the data are just a noisy version of the target image, with changes based on whether the image is blurred as well. Information from nearby sites might be combined or superimposed as a result of blurring. In these situations, an image of the data frequently matches the goal image and, in certain cases, can be helpfully interpreted by knowledgeable professionals without requiring additional processing [5]. In these situations, image reconstruction is employed to eliminate or at least lessen the impacts of blur and noise, improving the accuracy of any quantitative summary and the reliability of picture interpretation. Although noise in the data-gathering process is likely to be a part of the second kind of imaging problem, the primary focus here is a projective transformation, similar to that of tomography. Unlike the previously discussed blurring, projection has the effect of mixing, or superimposing, data from physically distant sites. This makes it considerably harder, if not impossible, to read the data directly; hence, it is crucial to undo the projection and do denoising and deblurring. It is true that image reconstruction in biomedical imaging is mathematically based and revolves around resolving forward and inverse difficulties. These procedures convert unprocessed, frequently imprecise measurements into insightful pictures that show the composition or operation of biological systems.

3.1.2 Forward and Inverse Problems in Image Reconstruction

Image reconstruction in biomedical imaging is fundamentally a mathematical process that involves solving forward and inverse problems.

3.1.2.1 Forward Problems

Modeling the way the physical system produces quantifiable data from the item being imaged is known as the forward problem in image reconstruction. It explains how the mechanics of the imaging modality convert an object's intrinsic characteristics—such as tissue density or conductivity—into detectable signals. For example, MRI describes how magnetic field gradients encode spatial information into frequency-domain signals, and CT represents how X-rays attenuate as they travel through tissues. To solve the inverse problem and retrieve the image, the forward problem is mathematically represented as a system operator applied to the object, frequently with additional noise. Accurate modeling of the forward problem in image reconstruction requires addressing several challenges. First, the complexity of physical phenomena must be considered, as the imaging

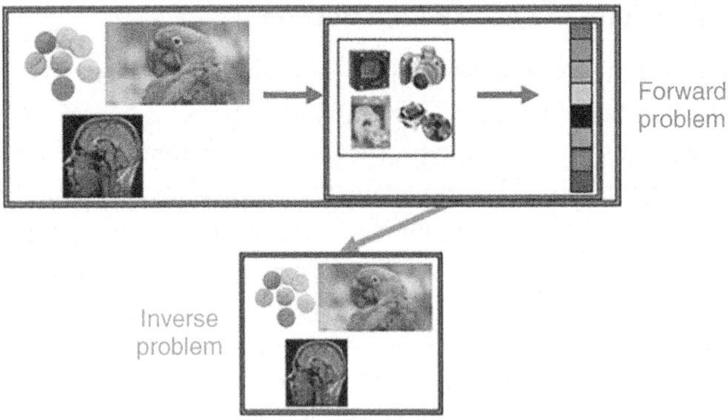

Figure 3.1 Forward and inverse problems in image reconstruction [6].

process often involves intricate physics such as multiple scattering, wave interference, or non-linear effects, all of which need to be accurately captured in the model. Additionally, measurement noise and system imperfections introduce uncertainties that can affect the quality of the data and must be incorporated into the forward model to ensure robustness. Finally, forward problem computations can be computationally demanding, particularly for high-resolution images or complex systems, necessitating significant computational resources and optimization techniques. Figure 3.1 demonstrates forward and inverse problems in image reconstruction.

For example:

- In X-ray CT, the forward model describes the attenuation of X-ray beams as they pass through different tissues.
- In PET, the forward problem models the detection of gamma photons emitted from radiotracers in the body.

Forward problems are typically well-posed, meaning they have a unique solution that depends continuously on the input data.

3.1.2.2 Inverse Problems

Recovering an object's interior structure or characteristics (such as tissue density or molecule concentration) from the information obtained by an imaging device is known as the "inverse problem" in image reconstruction. The inverse problem aims to reconstruct the object given the data, as opposed to the forward problem, which predicts the data given the object. Due to the ill-posed character of many inverse problems—where solutions might not exist, be unique, or depend steadily on the data—this approach is frequently mathematically difficult [7].

For instance:

- In MRI, reconstructing an image involves solving an inverse Fourier transform of the acquired frequency domain data.
- In CT, reconstructing a cross-sectional image from projection data requires solving an inverse Radon transform.

3.1.2.2.1 Challenges in Inverse Problems Inverse problems are ill-posed when one or more of the following conditions are not met:

1) **Existence:** A solution may not exist for the given data.
2) **Uniqueness:** There may be multiple solutions consistent with the measured data.
3) **Stability:** Small changes in input data can lead to large variations in the solution, making it sensitive to noise.

To address these challenges, regularization techniques are often employed. Regularization imposes additional constraints or incorporates prior knowledge to stabilize the solution. For example, Tikhonov regularization minimizes the influence of noise, and total variation regularization is used to preserve edges in images while reducing noise [8].

3.2 Classical Analytical Methods

Classical analytical methods for image reconstruction leverage direct mathematical transformations to solve inverse problems. These methods are computationally efficient and were among the first to be developed for biomedical imaging.

3.2.1 Filtered Back Projection (FBP)

Filtered back projection (FBP) is a widely used analytical method, particularly in CT and single-photon emission computed tomography (SPECT). It reconstructs images by combining projection data acquired at multiple angles (Figure 3.2) [9, 10]. An example of how the filtered back-projection (FBP) and back-projection reconstruction techniques work. A basic Gaussian ring distribution and its associated RT are displayed in the top panel. The distribution is reconstructed in the lower panel using (left to right) 5, 10, and 50 projections using the BP and FBP iRT algorithms. It is evident that the BP method results in weak and hazy reconstructions, even for a large number of projections. However, if sufficient projections are applied, the FBP algorithm can yield high-quality results.

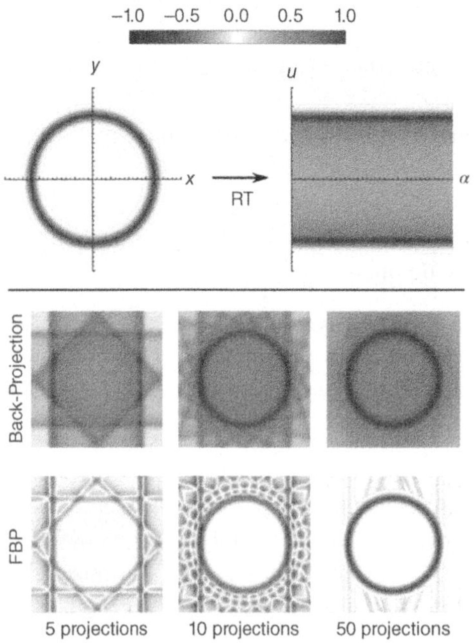

Figure 3.2 Filter back projection workflow.

- **Process:**
  1) Apply a mathematical filter to projection data in the frequency domain to correct for blurring.
  2) Back-project the filtered data across the image domain to reconstruct the image.
- **Advantages:**
  - Fast and computationally efficient.
  - Straightforward to implement.
- **Limitations:**
  - Highly sensitive to noise and data incompleteness.
  - Results in artifacts, such as streaks, when data quality is poor or when projections are undersampled.

3.2.2 Fourier-Based Methods

Fourier-based methods rely on the Fourier transform, which represents a signal in the frequency domain. These methods are particularly relevant in MRI, where raw data are acquired directly in the Fourier domain (k-space) [11].

- **Process in MRI:**
  1) Raw k-space data are acquired via magnetic field gradients.

2) An inverse Fourier transform is applied to reconstruct the spatial-domain image.

- **Applications:**
  - o Used extensively in MRI and optical imaging.
  - o Facilitates advanced techniques such as parallel imaging and CS.
- **Challenges:**
  - o Sensitive to noise and motion artifacts.
  - o Requires complete data acquisition for accurate reconstruction, which can increase scan times [12].

3.2.3 Algebraic and Iterative Techniques

Algebraic and iterative reconstruction methods take a different approach by framing the reconstruction problem as a system of linear equations. These methods are flexible and robust to incomplete or noisy data, making them suitable for challenging imaging scenarios.

3.2.3.1 Algebraic Reconstruction Techniques (ARTs)

Algebraic reconstruction technique (ART) is an iterative method that reconstructs images by solving a set of linear equations derived from the imaging system.

- **Process:**
  1) Represent the image as a vector of unknowns (pixel or voxel intensities).
  2) Solve the linear system iteratively, updating the image estimate by projecting measured data onto the solution space.
- **Advantages:**
  - o Handles incomplete or noisy data effectively.
  - o Can incorporate prior knowledge or regularization into the reconstruction process.
- **Limitations:**
  - o Computationally intensive compared to analytical methods.
  - o Convergence can be slow, depending on the problem size and constraints.

3.2.3.2 Simultaneous Algebraic Reconstruction Technique (SART)

SART improves upon ART by simultaneously updating all elements of the image vector at each iteration, rather than sequentially updating one equation at a time [13].

- **Process:**
  1) Each iteration considers all projection data, distributing the error correction uniformly across the image.
  2) Updates are smoothed to minimize oscillations and improve convergence.

- **Applications:**
  1) Widely used in CT and PET for iterative reconstruction.
  2) Suitable for sparse or undersampled data scenarios.
- **Advantages:**
  1) Faster convergence than ART.
  2) Produces smoother images with fewer artifacts.
- **Challenges:**
  1) Still computationally demanding for large-scale problems.
  2) Requires careful tuning of regularization parameters to balance noise suppression and image detail.

3.3 Convergence and Computational Challenges

Iterative techniques often require multiple iterations to achieve convergence, where the reconstructed image closely matches the measured data while satisfying imposed constraints. Convergence speed depends on factors such as:

- The choice of algorithm and optimization strategy.
- The quality and completeness of input data.
- Computational resources available for reconstruction.

To address these challenges, advancements in parallel computing and hardware acceleration (e.g., GPUs) have been integrated into iterative reconstruction pipelines. Additionally, hybrid approaches that combine iterative methods with deep learning are emerging as powerful tools to reduce computational burden while maintaining image quality (Table 3.1) [14, 15].

3.4 Signal Processing for Noise and Artifact Management

Modern medical research and diagnosis rely heavily on biomedical imaging, which offers vital insights into the anatomy and physiology of the human body. However, noise and artifacts—unwanted elements that obscure or alter the target structure's genuine representation—often impair imaging systems' efficacy. Artifacts are systematic distortions that result from technical, computational, or physical constraints in the imaging system, whereas noise is defined as random variations or fluctuations in the image that are usually introduced during the data-collecting process. Both of these elements deteriorate image quality, lower contrast and resolution, and make it difficult to derive useful information for

Table 3.1 Convergence and computational challenges.

Section	Topic	Details	Examples/Applications
2.1 Forward and Inverse Problems	Forward problems	Models data acquisition using physical laws	CT: X-ray attenuation; PET: Gamma detection
	Inverse problems	Recovers the object from measured data	MRI: Inverse Fourier; CT: Inverse Radon
	Challenges	Ill-posed problems with noise and instability	Regularization methods like Tikhonov
2.2 Classical Analytical Methods	FBP	Filters and back-projects projection data	Standard in CT and SPECT imaging
	Fourier-based methods	Processes data in the frequency domain	MRI reconstruction from k-space data
2.3 Algebraic and Iterative Techniques	ART	Iterative linear equation solving	Effective for noisy or sparse data (CT, PET)
	SART	Simultaneous updates for faster convergence	Preferred in sparse CT reconstruction
	Challenges	Computationally intensive, slow convergence	Accelerated with GPUs and hybrid ML methods

research or diagnosis. Advanced signal processing techniques are essential for enhancing image clarity and dependability to overcome these obstacles. Without sacrificing the underlying data, these methods seek to reduce noise, eliminate artifacts, and improve the visibility of essential features. To ensure that the reconstructed images stay true to the original signal, denoising methods, for instance, use mathematical models to suppress random noise while maintaining structural details. Contrarily, artifact correction techniques are intended to detect and correct systematic mistakes that arise from constraints in data capture or reconstruction procedures, such as motion blur, susceptibility artifacts in MRI, or streak artifacts in CT. Various imaging techniques have distinct sources of noise and artifacts. While thermal noise and artifacts from patient movement or hardware flaws can affect MRI, quantum noise from the small number of photons that reach the detector is a typical problem in CT and X-ray imaging. Scattering and absorption-related errors are common in optical imaging systems, and because sound waves are coherent, speckle noise poses a problem for ultrasound imaging. Researchers and clinicians can adjust noise reduction and artifact correction techniques to the unique features of each modality by being aware of these sources. The physical and technical mechanisms underlying noise and artifacts are examined in this section. Additionally, it looks at a variety of denoising techniques that are becoming more and more popular due to their capacity to identify intricate patterns in data, including deep learning-based methods, wavelet transform techniques, and spatial filtering. Furthermore, a thorough discussion of artifact correction strategies is provided, emphasizing their applications in a variety of imaging modalities. These techniques include model-based approaches, iterative reconstruction algorithms, and machine-learning techniques. When combined, these developments in signal processing enable researchers and doctors to acquire dependable, high-quality images, which eventually enhance diagnostic precision and advance biological research [16].

3.4.1 Sources of Noise and Artifacts

In biomedical imaging, noise and artifacts are common problems that have a substantial impact on image quality and diagnostic precision. The intrinsic constraints of imaging technologies, patient mobility during data collection, and outside variables such as equipment instability or environmental influence are some of the causes of these problems. Usually appearing as random fluctuations in the image, noise is frequently created by the imaging device itself. For example, thermal noise in MRIs from electrical components or quantum noise in X-rays and CT scans due to low photon counts [17]. On the other side, artifacts are systematic distortions that might result from things like reconstruction errors, hardware flaws, or motion during imaging. Susceptibility artifacts in MRIs, shadowing or speckle in ultrasound pictures, and streak artifacts in CT scans are a few examples. Improving imaging results requires an understanding of the causes of noise and artifacts. By determining their origins, scientists and medical professionals can create focused mitigation plans that guarantee the recreated pictures accurately depict the biological structures or processes under investigation. For example, sophisticated motion correction algorithms can be used to reduce related artifacts in dynamic imaging situations when patient movement is unavoidable. In a similar vein, noise reduction methods like adaptive filtering or signal averaging can assist in reducing random fluctuations while maintaining crucial structural elements. Furthermore, outside variables such as power supply fluctuations or electromagnetic interference can add more complexity, highlighting the necessity of strong environmental control and system calibration. Imaging specialists can overcome these difficulties and push the limits of what is possible with existing imaging technologies, ultimately improving their reliability and diagnostic potential, by carefully examining the operational, technical, and physical factors that contribute to noise and artifacts. Figure 3.3 clearly depicts the difference between noise and artifacts.

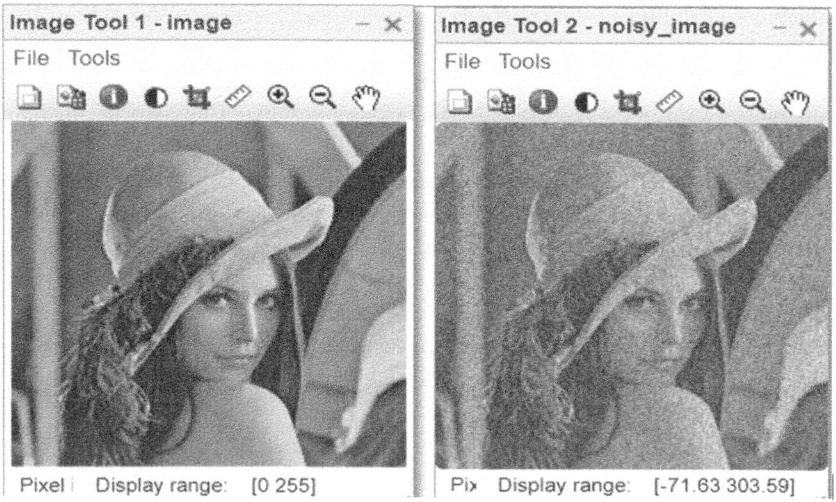

Figure 3.3 Noise vs artifacts.

3.4.2 Sources of Noise

Noise refers to random variations in the acquired data that obscure the true signal. Common types of noise include:

- **Gaussian Noise:**
  One kind of random noise that is frequently seen in biomedical imaging systems is Gaussian noise, sometimes known as white noise. It is distinguished by its statistical characteristics, in which the noise intensities have a normal distribution with a particular standard deviation defining its spread and a mean value, usually zero, at its center. Its homogeneous power distribution across all frequencies—which is independent of the frequency content of the signal—gives it the "white" name. A basic drawback of many imaging systems, this kind of noise can seriously impair image quality by hiding minute features or creating ambiguities in signal interpretation. Gaussian noise is a frequent problem in MRI, mostly due to thermal fluctuations in the electrical parts of the system, including the amplifiers and receiver coils. During the data acquisition process, these oscillations cause random variations in the signal, especially when imaging weak signals or at low field strengths. Similar to this, Gaussian noise—also known as Johnson–Nyquist noise—is introduced into optical imaging systems by thermal vibrations in the electronic circuits of the sensor. In low light, when the signal-to-noise ratio (SNR) is already impaired, it becomes especially noticeable. Gaussian noise has two effects on imaging: it lowers the SNR and can mask fine details that are important for study or

diagnosis. Brighter or darker pixels are sporadically scattered throughout the image, giving the appearance of a grainy or speckled texture. Because of its existence, denoising procedures are required to enhance image clarity and diagnostic usefulness. Spatial filtering techniques like Gaussian or median filters, which smooth out random variations while maintaining edges, and more sophisticated methods like wavelet-based denoising, which uses multi-scale decomposition to selectively remove noise without obscuring crucial details, are common ways to reduce Gaussian noise. Developing specialized noise reduction techniques requires an understanding of the origin and properties of Gaussian noise. Imaging systems can attain more fidelity and enhanced reliability by resolving this prevalent constraint, which would allow researchers and doctors to derive more significant insights from the collected data [18].

- **Poisson Noise:**
  A basic kind of statistical noise, Poisson noise, usually referred to as shot noise, occurs in photon-counting systems frequently employed in biomedical imaging modalities such as PET and CT. The discontinuous and stochastic nature of photon emission and detection processes is intrinsically linked to this noise. Poisson noise has a variance equal to its mean, which means that its intensity is directly proportional to the quantity of photons detected, in contrast to Gaussian noise, which has a uniform variance. Because of this characteristic, Poisson noise is especially problematic in low-photon count situations, like low-dose imaging. For instance, in CT imaging, X-ray photons travel through the body and are attenuated according to the different tissue densities. The quantity of photons that reach detectors following this attenuation is measured. A higher degree of Poisson noise is produced in low-dose CT because fewer X-ray photons are released to lower the radiation exposure to the patient. This may show up in the reconstructed images as mottled or grainy textures, which could obscure fine anatomical details and lower diagnostic precision. Similar to this, Poisson noise becomes important in PET, where gamma photons are detected after radioactive tracers decay, when tracer concentrations are low or acquisition times are brief, making it more difficult to precisely reconstruct tracer distributions. Because Poisson noise directly affects the SNR and overall image quality, its effects are especially significant in clinical and research situations. Advanced denoising and reconstruction algorithms are used to lessen its impact. Noise can be effectively suppressed while maintaining the integrity of the signal using methods like statistical modeling, which uses the Poisson distribution in the reconstruction process. In PET, iterative reconstruction techniques that model the noise characteristics to improve image quality, including maximum likelihood expectation maximization (MLEM) or Bayesian approaches, are very successful [9].

- **Thermal Noise:**
  The random migration of electrons in conductive materials, like the detectors used in imaging systems, produces thermal noise, often referred to as Johnson–Nyquist noise, a type of electronic noise. This noise is a natural byproduct of the system's thermal energy, where temperature causes variations in the passage of electrons within the material. As the temperature rises, thermal noise becomes more noticeable since its intensity is directly correlated with both the material's resistance and temperature. In imaging modalities like ultrasound and optical imaging, where sensitive detectors are utilized to collect weak signals, frequently in low-intensity or low-light settings, this kind of noise is very significant. Thermal noise occurs in the piezoelectric transducers used in ultrasound imaging, which transform sound waves into electrical impulses. Although thermal noise can mask weak signals and reduce image quality, especially in low-intensity scans or deep tissue imaging, these transducers are extremely sensitive to even slight pressure changes. While the signal strength is already low, as it is during high-frequency ultrasonic imaging or while scanning through dense tissues, the noise becomes more noticeable. Similar to this, thermal noise can be produced in photodetectors used in optical imaging, especially when imaging in low light or near-infrared. Several techniques are used to lessen the impact of thermal noise. Techniques including signal averaging and the use of more powerful transmitters can assist increase the SNR and lessen the relative influence of noise in ultrasound. Furthermore, to improve the intended signal and reduce thermal noise, filtering techniques can be applied, both software-based (like adaptive filtering) and hardware-based (like low-noise amplifiers). Cooling systems can be used in optical imaging to lower detector temperatures and lessen thermal noise, especially in low-light imaging applications like fluorescence microscopy and optical coherence tomography (OCT). To improve image quality by lowering noise without changing the underlying structures, sophisticated image processing algorithms like wavelet transforms and denoising filters are also used [19].

3.4.3 Sources of Artifacts

Artifacts are systematic distortions or errors in the reconstructed image caused by the imaging process or patient-related factors. Common artifacts include:

- **Motion Artifacts:**
  Involuntary patient movement during picture acquisition is one of the main sources of motion artifacts. For instance, the image may become blurry or distorted due to breathing, heartbeat, or even little patient pain. Even slight motions might cause the scanned data to become misaligned, leading to aberrations like ghosting or blurring in MRIs, which capture high-resolution pictures

over a comparatively lengthy period. Even while patients are usually told to stay motionless during imaging, voluntary motions like talking, chewing, or changing postures can sometimes result in motion artifacts, especially when using modalities like CT, MRI, and ultrasound [20].

- **Metal Artifacts:**
  When metal objects like implants, dental fillings, prosthetics, or surgical clips are present close to the region of interest, metal artifacts—common and difficult distortions in medical imaging—occur. This is especially true for techniques like CT, MRI, and X-ray imaging. The metal's interaction with the imaging equipment results in these artifacts, which distort or imprecise the image. Metal objects in CT create regions of high attenuation by absorbing X-ray photons more efficiently than the surrounding tissues. This leads to beam hardening and scatter, which produces distorted pictures and streak aberrations. Local magnetic field inhomogeneities caused by metal implants in MRI result in signal voids and geometric distortions in the surrounding tissue, which reduce signal strength and make it harder to see anatomical structures. Metal objects reflect and refract sound waves in ultrasonography, creating shadowing artifacts that give the space behind the metal a distorted and empty appearance. While MRI utilizes specialist metal artifact reduction sequences (MARS), spatial saturation techniques, and non-metallic implants to reduce these artifacts, CT scanners use artifact reduction algorithms, dual-energy CT, and higher radiation doses. When metal-induced shadowing in ultrasound dramatically alters the image, doctors may change the probe angle or employ other diagnostic modalities like CT or MRI. In the presence of metallic objects, enhancing image quality and guaranteeing precise diagnosis and treatment planning depend on an understanding of and attention to metal artifacts [21].
- **Beam-Hardening Artifacts:**
  Hardening of Beams: The differential attenuation of X-ray beams when they travel through materials with differing densities, such as bone, metal implants, or contrast chemicals, causes artifacts, which are frequent distortions in CT imaging. When X-rays travel through dense materials, they are absorbed more readily by lower-energy (softer) X-rays than by higher-energy (harder) X-rays. Consequently, the image reconstruction is distorted due to the increased proportion of high-energy X-rays in the emerging X-ray beam. A wide range of X-ray energies are employed in CT imaging to provide finely detailed cross-sectional pictures of the body. Higher-energy photons are more easily absorbed, while lower-energy photons are more readily absorbed when the X-ray beam comes into contact with dense objects like metal or bone. The X-ray beam becomes "harder" or more energetic as a result, creating an imbalance in energy levels that generates distortions like bands or dark streaks. These artifacts may appear as cupping artifacts in regions of dense bone or tissue,

or as black bands or streaks emanating from metal items such as implants or surgical clips. Beam-hardening artifacts can be made worse by the use of contrast materials, high-density materials, and poor calibration. Modern CT scanners include metal artifact reduction (MAR) techniques, dual-energy CT, artifact reduction algorithms, and higher scan doses to lessen these effects. Beam hardening can also be lessened by optimizing scan parameters such as tube current, voltage, and scan angles. Notwithstanding the difficulties these artifacts provide, improvements in CT technology have made it possible to lessen these distortions, giving doctors clearer, more precise images for enhanced diagnosis and treatment planning. In CT imaging, beam hardening occurs when lower-energy X-rays are absorbed more readily than higher-energy X-rays, leading to cupping or streaking artifacts. Addressing these noise and artifact sources is essential for ensuring diagnostic accuracy and minimizing false positives or negatives in clinical settings [22].

3.5 Denoising Techniques

Denoising techniques aim to suppress noise while preserving important image details. These techniques are broadly classified into spatial domain filtering and transform domain approaches.

3.5.1 Spatial Domain Filtering

Spatial domain methods operate directly on pixel intensities, applying filters to smooth noise and enhance image quality.

- **Median Filtering:**
  Median filtering replaces each pixel value with the median of its neighborhood, effectively removing salt-and-pepper noise. It is widely used in ultrasound and optical imaging.
  - **Advantages:** Simple and effective for impulsive noise.
  - **Limitations:** May blur fine details in high-resolution images.
- **Gaussian Filtering:**
  Gaussian filters use a Gaussian kernel to perform weighted averaging of neighboring pixels, reducing high-frequency noise.
  - **Applications:** Commonly applied in MRI and CT imaging for smoothing.
  - **Limitations:** Can blur edges and reduce image sharpness.

3.5.2 Transform Domain Approaches

Transform domain methods process images in the frequency or wavelet domain to selectively attenuate noise components.

- **Wavelet-Based Denoising:**
  Wavelet transforms decompose the image into multiple scales, isolating noise in high-frequency components. Thresholding techniques (e.g., hard or soft thresholding) are then applied to suppress noise.
  - **Advantages**: Preserves edges and fine details.
  - **Applications**: Widely used in MRI, CT, and optical imaging.
- **Fourier-Based Denoising:**
  Fourier transforms convert the image into the frequency domain, where low-pass filters remove high-frequency noise.
  - **Applications:** Useful for periodic noise removal in modalities like ultrasound.
  - **Limitations:** Struggles with non-stationary or spatially localized noise.

These techniques form the foundation of noise suppression, with the choice of method depending on the imaging modality, noise characteristics, and clinical requirements [23].

3.6 Artifact Correction Methods

Artifacts pose significant challenges to image quality, requiring targeted correction techniques that address their specific causes.

3.6.1 Model-Based Correction Techniques

Model-based approaches use mathematical and physical models of the imaging process to correct artifacts.

- **Motion Correction:**
  Motion artifacts are addressed by realigning the acquired data or incorporating motion models into the reconstruction process.
  - **Applications:**
    In MRI, retrospective gating techniques compensate for cardiac or respiratory motion.
    In ultrasound, real-time tracking systems minimize motion effects.
- **Beam-Hardening Correction:**
  Beam-hardening artifacts in CT are mitigated by applying energy correction models or dual-energy scanning techniques that differentiate low- and high-energy X-rays.
- **Metal Artifact Reduction (MAR):**
  MAR algorithms correct streaking caused by metal implants by interpolating or re-weighting affected projection data.

- o **Techniques:** Projection completion, iterative reconstruction, or hybrid approaches that combine both.

3.6.2 Deep Learning Approaches for Artifact Reduction

Recent advances in AI have led to the development of deep learning methods for artifact correction. These approaches leverage neural networks to learn and correct complex artifacts from large datasets [24].

- **Convolutional Neural Networks (CNNs):**
  CNNs are trained on paired noisy and clean images to predict artifact-free reconstructions.
  - o **Applications:**
    - MRI: Corrects motion and undersampling artifacts.
    - CT: Reduces beam-hardening and metal artifacts.
- **Generative Adversarial Networks (GANs):**
  GANs generate high-quality reconstructions by learning the distribution of artifact-free images.
  - o **Advantages:** Superior performance in preserving image details.
  - o **Limitations:** Computationally intensive and reliant on large datasets.
- **Hybrid Approaches:**
  Combining model-based and deep learning methods integrates the strengths of both approaches, offering improved accuracy and robustness. For instance, physics-informed neural networks incorporate imaging system models into the learning process (Table 3.2) [25].

3.6.3 Advanced Signal Processing Strategies

Signal processing has evolved significantly over the past few decades, incorporating a wide range of techniques designed to efficiently capture, analyze, and reconstruct signals. These techniques are crucial in numerous applications, especially in fields like medical imaging, telecommunications, and multimedia.

Table 3.2 Accuracy and performance comparison.

Method	PSNR (dB)	SSIM	Training time	Hardware requirements	Best for
CNN	28–32	0.90–0.94	Fast	Moderate GPU	General artifact removal
GAN	30–34	0.92–0.96	Slow	High-end GPU	Detail preservation
Hybrid	32–36	0.94–0.98	Medium	Moderate GPU	Physics-based artifacts

Among the advanced strategies, three of the most prominent are CS, model-based reconstruction techniques, and the application of machine learning and neural networks in signal processing. Each of these approaches contributes to improving the performance of signal reconstruction, especially when data acquisition is sparse or when dealing with noisy or incomplete data [26].

3.7 Compressed Sensing in Imaging

CS is a revolutionary concept in signal processing that allows for the reconstruction of signals from fewer samples than traditionally required. In conventional signal processing, the Nyquist–Shannon sampling theorem stipulates that signals must be sampled at a rate at least twice the highest frequency component of the signal. However, in many applications, the signal of interest is sparse or compressible in some basis or transform domain, which means it can be represented with far fewer coefficients than would be needed in the time or spatial domain [27].

3.7.1 Sparse Representation and Sampling

The concept of sparsity is central to CS. A signal is considered sparse if most of its coefficients (in a suitable transform domain) are zero or near zero. For example, natural images, audio signals, and medical imaging data often exhibit sparsity when transformed into domains like wavelets, Fourier, or discrete cosine transforms (DCT). The primary idea of CS is that instead of sampling a signal at a high rate in the original domain, we can sample it in a manner that directly captures its sparse features [28].

In CS, the number of measurements or samples is much smaller than that suggested by the Nyquist criterion, but provided that the sampling method is designed properly, the signal can still be reconstructed accurately. This is made possible by leveraging two principles: the sparsity of the signal and the incoherence of the sampling basis. The sampling process involves multiplying the signal by a measurement matrix, which is designed to be incoherent with the basis in which the signal is sparse.

Using optimization algorithms, such as L1-norm minimization (which can be solved via techniques like basis pursuit), it is possible to recover the sparse signal xxx from the measurement yyy.

3.7.2 Applications in MRI and CT

In medical imaging, MRI and CT are two of the most common applications where CS is applied. These imaging techniques typically involve acquiring a large number of measurements over a long period, leading to issues such as high scan time, motion artifacts, and patient discomfort. By utilizing CS, it is possible to significantly reduce the number of samples needed for high-quality imaging, thus improving efficiency and reducing acquisition time.

- **MRI:** In MRI, CS allows for faster acquisition of images without compromising on the resolution or accuracy of the images. It relies on the fact that MRI images often exhibit sparsity in the wavelet or Fourier domain. By sampling fewer k-space points (frequency domain), CS algorithms reconstruct the image through iterative techniques, reducing scan time by up to 50–70%.
- **CT:** In CT imaging, CS helps reduce radiation exposure to patients by requiring fewer projections to reconstruct high-quality images. Traditional CT scans require a large number of X-ray images from different angles. With CS, fewer projections are needed, and the image can be reconstructed with minimal loss of detail.

These applications exemplify how CS reduces computational time and resources, benefiting clinical settings where speed and accuracy are paramount.

3.7.3 Model-Based Reconstruction Techniques

While CS relies heavily on sparse representations, model-based reconstruction techniques emphasize incorporating prior knowledge about the physical model of the imaging process. These models are designed to account for various aspects of the system, such as sensor noise, motion artifacts, and geometric distortions, to enhance the reconstruction of the original signal or image.

3.7.4 Bayesian Inference Models

One of the most popular approaches in model-based reconstruction is Bayesian inference. In Bayesian methods, reconstruction is framed as a probabilistic problem. The goal is to estimate the most likely signal or image.

Bayesian methods allow for incorporating prior knowledge about the signal. For example, in medical imaging, prior knowledge might include the sparsity of the image, the expected smoothness of anatomical structures, or the spatial correlation between different regions of the image. This is particularly useful when the measurements are noisy or incomplete, as Bayesian techniques can incorporate noise models into the reconstruction process.

Bayesian models are typically solved using Markov Chain Monte Carlo (MCMC) methods or other sampling techniques to estimate the posterior distribution. This results in more accurate and robust reconstructions, especially in low-dose imaging applications, such as low-dose CT scans.

3.8 Statistical Methods for Noise Modeling

In addition to Bayesian methods, statistical approaches are frequently used to model noise in imaging systems. These models help account for the randomness and distortions introduced by the acquisition hardware, environmental factors, or physiological motion. In medical imaging, for instance, noise may arise from various sources such as thermal noise in detectors, quantization noise, and motion artifacts.

A common technique used to model noise is Gaussian noise modeling. However, in many real-world applications, especially in medical imaging, the noise is often not Gaussian and can follow distributions such as Poisson noise (commonly observed in X-ray imaging) or Rician noise (common in MRI).

Statistical methods help to adapt the reconstruction process to these non-Gaussian noise distributions, ensuring more accurate images.

3.8.1 Machine Learning and Neural Networks

Machine learning has made significant strides in signal processing, and it is increasingly being applied to both CS and model-based reconstruction techniques. Specifically, neural networks and deep learning models have shown remarkable success in tasks like denoising, artifact removal, and image reconstruction.

3.8.2 Supervised vs Unsupervised Approaches

Machine learning in signal processing can be broadly categorized into supervised and unsupervised approaches.

- **Supervised learning** involves training a model on labeled data, where both the input and the corresponding target output are known. For example, in medical imaging, supervised models can be trained to map noisy or incomplete MRI data to high-quality reconstructions. The network learns a mapping function that optimizes the reconstruction error, minimizing artifacts while preserving essential features of the image.
- **Unsupervised learning**, on the other hand, does not require labeled data. Instead, the model learns patterns in the data through methods such as clustering, autoencoders, or generative models. For example, unsupervised methods can learn to remove artifacts from noisy signals by leveraging the inherent structure or properties of the data, such as smoothness or sparsity.

3.8.3 Deep Learning for Artifact Removal and Reconstruction

Deep learning techniques, particularly convolutional neural networks (CNNs), have been applied extensively to signal processing tasks like artifact removal, denoising, and image reconstruction. Deep learning models can automatically learn hierarchical features of data, making them highly effective in reconstructing signals from sparse or noisy measurements.

For example, deep CNNs have been trained to reconstruct high-quality images from noisy or undersampled MRI data. The network learns to map the noisy input to the clean output by minimizing the reconstruction error through training on large datasets of high-quality MRI scans. Once trained, the model can then generalize to new data, enabling fast and efficient image reconstruction from sparse measurements.

Another exciting development is the use of generative adversarial networks (GANs) for signal reconstruction. GANs consist of two networks—a generator and a discriminator—that compete against each other. The generator produces reconstructed images, while the discriminator evaluates their quality. This adversarial training process results in the generator learning to produce highly realistic reconstructions that are difficult to distinguish from real, high-quality images.

Deep learning models can significantly accelerate the reconstruction process, enabling real-time imaging and enhancing the quality of reconstructed images. These techniques are particularly beneficial for applications where high computational efficiency and accuracy are critical, such as in real-time medical imaging.

3.8.4 Emerging Innovations in Image Reconstruction

Reconstruction of images is needed in many types of imaging systems, for instance, medical and diagnostic imaging systems, space-based imaging systems, and industrial imaging systems. Progress in technology has enhanced better solutions in computational methods in image reconstruction rendering it a central role in imaging. In this context, new approaches like hybrid computational techniques, multimodal and multiresolution techniques, quantum signal processing, and AI-based real-time reconstruction are hence opening up toward better solutions for image reconstruction systems. These developments have the potential of revolutionizing how imaging data is processed and analyzed with broad applications across diverse sectors ranging from medicine, and environmental management, to the development of autonomous systems.

3.9 Hybrid Computational Methods

The proposed approach is a blend between classical approaches of general image reconstruction and deep learning methods. Traditional approaches such as FBP or iterative reconstruction use mathematical algorithmic formulas for reconstructing images from raw data. These techniques have been the norm in imaging systems such as computer tomography or MRI for years. But there are issues when it comes to using such techniques such as in dealing with noisy data, in trying to solve for shorter time, and in improving the quality of images [29].

Compared to these, deep learning has been proven as an effective tool to address these limitations. CNNs and most of the other deep learning structures are especially effective at applying patterns such as image de-noising, eradicating artifacts, and increasing the resolution of an image. However, deep learning-based methods require a vast set of labeled training data and computational resources which can be very difficult in certain application cases.

It is shocking to note that an integration of the classical methods entwined with deep learning is an attempt to merge what is best. Hybrid methods are a combination of the mathematical rigor of classical methods and the learnability of deep learning, and it creates a result that is both better or faster than both these categories of approaches.

3.9.1 Optimization-Based Deep Networks

Optimization-based deep networks are one of the hybrid methods' subcategories, specifically concentrating on introducing optimization into deep learning that enhances image reconstructions. The standard practices for optimization toward image reconstruction have followed iterative algorithm optimization with a view of minimizing cost functions that afford comparison between the reconstructed image and the measured data. This framework presented can then be applied with deep learning to make optimization-based networks capable of learning these optimal features and representations good for different types of images and depths of noise.

The learned regularizers in CS represent one example of optimization-based deep networks in image reconstruction. In this case, deep neural networks are learned to serve as prior models that will enhance the recovery of image content from the sparse data which is a familiar problem in MRI and CT images. The statistical properties of images are discerned by the deep learning model, so it progressively adjusts the reconstructed representation, thereby mitigating the amount of noise.

3.9.2 Multimodal and Multiresolution Techniques

Multimodal image reconstruction usually covers the use of data produced by multiple modality imaging techniques to get a more defined look at the object or tissue. For instance, PET is used to obtain details about function, particularly with aspects relating to metabolism; on the other hand, MRI will provide detailed structural details of the tissues. Thus physicians can get more detailed information concerning a certain patient's status, thereby increasing the chance for correct identification of the disease and accurate planning of the further treatment process.

Integration of multimodal data is challenging because the information used by various modalities is collected in different ways, at different resolution levels, and may contain different kinds of information. The registration and alignment processes are crucial in multimodal image reconstruction because the data acquired from different imaging modalities should reflect the same features in the anatomy of the subject so that the images can be reconstructed into a whole. Latest image fusion techniques, which are usually backed by deep learning, have indicated the potentiality in enhancing the reconstruction of multimodal images for enhancing segmentation, classification, and interpretation of the fused data.

For instance, in the case of merging PET–MRI images, the machine learning models are helpful regarding issues to do with alignment and clearing of differences in the manifestation of the images. Such models could also aid in identifying related features of different modalities such as functional and anatomical features that may be useful in clinical decision making. The potential of fusing functional and structural information is the most promising for cancer imaging, functional brain MRI and MRS, as well as cardiac imaging.

3.9.3 Super-Resolution Approaches for Enhanced Detail

SR stands for super-resolution which allows for achieving better definition, with the intent of recovering details that otherwise cannot be seen. From the image reconstruction point of view, SR techniques are especially valuable if the imaging systems have low resolution. For instance, MRI scans of the organs or ultrasound images of the fetus may be blurry because of some technical conditions or lack of advanced gear. Enhancement methods, frequently based on deep learning, try to obtain high-frequency details from low-frequency images.

These approaches cope by introducing a set of patterns derived from training a large number of high- resolution image samples and their respective low-quality counterparts. A deep neural network is then learnt to predict the details of the high-resolution image from a low-resolution one. This can greatly improve the diagnostic capability of the image since small lesions or minor anatomical details can be easily overlooked from a low-resolution image. Then in areas such as satellite imagery or remote sensing, SR may improve the quality of images obtained from drones or satellites and come up with an understanding of environmental status.

3.10 Quantum Signal Processing

Image reconstruction has been an area of interest in quantum computing for a long time despite the current limited developments in the field. Unlike traditional computing that deals with bits, they exist in a state of 0 and 1. Specifically, quantum computers operate on the back of quantum, or qubits that can exist in two or more states at one time because of a principle known as superposition. This property makes quantum computers possess better computational ability than classical computers; if incorporated into image reconstruction, it is likely to immensely improve the efficiency of the process. It revealed that quantum computing has promises of application in the field of image reconstruction for optimizing algorithms of accelerating, practicing image segmentation, and improving computational speed. For instance, biological imaging allows for decreasing the time needed for the reconstruction of images from large sets of data down to real time, potentially revolutionizing medical imaging and remote sensing. A quantum signal processing unit (QSP unit) is a specific field within a quantum computing unit, which aims to use a quantum system for signal processing applications. There are implications of using QSP to improve the situation in the case of image reconstruction, or when dealing with noisy or incomplete data. Some quantum algorithms might be more faithful to the problem statement than their classical counterparts in settings such as denoising, interpolation, and data compression and yield images of greater quality from noisy or sparse data.

3.10.1 Quantum Imaging and Sensing

Besides quantum computing, quantum imaging techniques are being made to improve the image reconstruction techniques directly. Compared to classical sensors, quantum sensors take advantage of quantum relations such as entanglement to improve the accuracy of the measurements to be made. For example, new quantum-enhanced computed tomography or MRI could enable higher-density imaging and higher sensitivity of the displayed reconstructions, in medical imaging.

Although there is no direct implementation of practical quantum imaging systems today, they can be used in medical diagnosis, material science, and environmental analysis with better performance than conventional imaging systems. Figure 3.4 demonstrates the difference between a quantum image and a classical image.

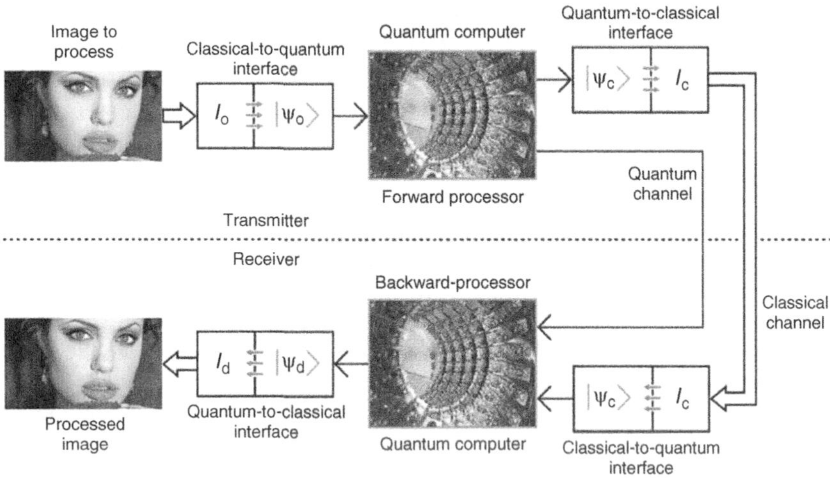

Figure 3.4 Quantum image vs classical image [30].

3.11 AI-Assisted Real-Time Reconstruction

Real-time image reconstruction is another application that is rapidly growing, especially in the point-of-care (POC) systems. For many medical imaging modalities, it is indispensable to obtain images as soon as possible and with high precision. However, the computational resources needed to generate high-quality images for subsequent use in the identification of serviceable pathological features are often unavailable or limited in centers such as rural clinics or in areas that are remote.

This challenge is being tackled by edge computing coupled with embedded AI that shall allow the reconstruction of images in the target devices in real time. Edge computing is the practice of performing data analysis at the endpoints, which can be a smartphone, tablet, medical imaging device, or any other device. This cuts the time needed in commands which also promotes faster decision-making processes.

Real-time reconstruction systems that make use of AI can tap into these devices by employing machine learning models. These models indeed are developed in such a way that they can make required image reconstructions from raw data in a relatively short time and with minimum computational power. For instance, in portable ultrasound systems, advanced AI models can be put into practice to help with real-time reconstructions of images and improve image quality, enabling healthcare decision-makers to make decisions in real time.

It also allows the automation of the data preprocessing stage, together with more complex image processing tasks, such as denoising, edge enhancement, and feature extraction to be performed on the frontline, at edge devices. This results in an improved ability to make accurate diagnoses, as opposed to complex diagnostic models requiring high-end computing environments to implement.

3.12 Conclusion

Biomedical image reconstruction is undergoing significant advancements as it increasingly incorporates computational methodologies, sophisticated signal processing techniques, and AI. The primary aim of this progression is to improve the precision, rapidity, and efficacy of reconstructing intricate medical images, thereby enhancing diagnostic and therapeutic capabilities. Conventional methodologies, such as Fourier-based techniques and analytical algorithms, continue to provide dependable foundations. Nevertheless, these approaches are now being supplemented by contemporary innovations, including CS, which facilitates accurate reconstruction from a reduced number of measurements, and Bayesian inference, which offers probabilistic frameworks for addressing uncertainty in imaging data. Novel paradigms, such as quantum image processing, are beginning to unveil new opportunities by harnessing the distinctive computational prowess of quantum systems.

Concurrently, the amalgamation of AI-driven methodologies, particularly deep learning and machine learning models, has revolutionized reconstruction workflows by autonomously discerning patterns and diminishing noise while expediting image generation. Edge computing further enhances these advancements by positioning computation in proximity to medical devices, thus enabling real-time image reconstruction and diagnostics. This is especially significant in remote or resource-constrained regions, where access to centralized computational infrastructure is frequently limited.

The convergence of these varied techniques highlights the intrinsically interdisciplinary essence of biomedical image reconstruction, drawing upon disciplines such as mathematics, physics, engineering, and computer science. Collectively, these advancements are not only refining the existing state of clinical imaging but also establishing the foundation for transformative changes in the imminent future. As innovation persists, medical imaging is set to become increasingly precise, accessible, and transformative for global healthcare.

References

1 Lang, L.F., Neumayer, S., Öktem, O., and Schönlieb, C.-B. (2020). Template-based image reconstruction from sparse tomographic data. *Applied Mathematics & Optimization* 82 (3): 1081–1109. https://doi.org/10.1007/s00245-019-09573-2.

2 Fu, J., Cong, P., Xu, S. et al. (2025). Neural architecture search with Deep Radon Prior for sparse-view CT image reconstruction. *Medical Physics* 52 (5): 3044–3058. https://doi.org/10.1002/mp.17685.

3 Wang, J., Shi, Y., and Liang, L. (2024). Progress in image recovery technology based on deep neural network. In: *International Conference on Electronics, Electrical and Information Engineering (ICEEIE 2024)* (ed. S. Li and B. Hu), 8. SPIE https://doi.org/10.1117/12.3052149.

4 Guan, Y., Li, Y., Liu, R. et al. (2023). Subspace model-assisted deep learning for improved image reconstruction. *IEEE Transactions on Medical Imaging* 42 (12): 3833–3846. https://doi.org/10.1109/TMI.2023.3313421.

5 Priyadharshini, L., Niranjana, K., Deepak Sakthivel, M.S., and Mohan, M.L. (2024). Pixel-level reconstruction of noisy underwater images using advanced deep learning models. *2024 4th Asian Conference on Innovation in Technology (ASIANCON)*, 1–5. https://doi.org/10.1109/ASIANCON62057.2024.10837921.

6 Monardo, V. J. (2022). Efficient algorithms for solving nonlinear inverse problems in image reconstruction, PhD dissertation, Department of Electrical & Computer Engineering, Carnegie Mellon University, Pittsburgh, PA.

7 Li, Y., Yang, H., Xie, D. et al. (2021). POCS-augmented CycleGAN for MR image reconstruction. *Applied Sciences* 12 (1): 114. https://doi.org/10.3390/app12010114.

8 Bostan, E., Kamilov, U.S., and Waller, L. (2018). Learning-based image reconstruction via parallel proximal algorithm. *IEEE Signal Processing Letters* 25 (7): 989–993. https://doi.org/10.1109/LSP.2018.2833812.

9 Reader, A.J., Corda, G., Mehranian, A. et al. (2021). Deep learning for PET image reconstruction. *IEEE Transactions on Radiation and Plasma Medical Sciences* 5 (1): 1–25. https://doi.org/10.1109/TRPMS.2020.3014786.

10 Sparling, C. and Townsend, D. (2022). Tomographic reconstruction techniques optimized for velocity-map imaging applications. *The Journal of Chemical Physics*, 157 (11). https://doi.org/10.1063/5.0101789.

11 Ye, H., Chen, Y., Hu, D. et al. (2023). Progress of deep learning based quantitative MRI reconstruction methods. *Journal of Image and Graphics* 28 (6): 1698–1708. https://doi.org/10.11834/jig.230041.

12 Bian, W. and Tamilselvam, Y.K. (2024). A review of optimization-based deep learning models for MRI reconstruction. *AppliedMath* 4 (3): 1098–1127. https://doi.org/10.3390/appliedmath4030059.

13 Ye, L. and Zou, C. (2023). A survey of image super-resolution reconstruction based on deep learning. In: *International Conference on Wireless Communications, Networking and Applications*, 584–594. Springer Nature Singapore https://doi.org/10.1007/978-981-99-3951-0_64.

14 Aghabiglou, A. and Eksioglu, E.M. (2021). MR image reconstruction based on densely connected residual generative adversarial network–DCR-GAN. In: *International Conference on Computational Collective Intelligence*, 679–689. Cham: Springer International Publishing https://doi.org/10.1007/978-3-030-88113-9_55.

15 Kak, A.C. and Slaney, M. (2001). Principles of Computerized Tomographic Imaging. *Society for Industrial and Applied Mathematics*. https://doi.org/10.1137/1.9780898719277.

16 Yau, H. and Du, X. (2021). Robust deep learning-based multi-image super-resolution using inpainting. *Journal of Electronic Imaging* 30 (01): 013005. https://doi.org/10.1117/1.JEI.30.1.013005.

17 Ben Yedder, H., Cardoen, B., and Hamarneh, G. (2021). Deep learning for biomedical image reconstruction: a survey. *Artificial Intelligence Review* 54 (1): 215–251. https://doi.org/10.1007/s10462-020-09861-2.

18 Pal, A. and Rathi, Y. (2022). A review and experimental evaluation of deep learning methods for MRI reconstruction. *Machine Learning for Biomedical Imaging* 1: 1–50. https://doi.org/10.59275/j.melba.2022-3g12.

19 Peng, Z., Jinchan, W., Huanqing, P. et al. (2024). Image compressed sensing reconstruction via deep image prior with feature space and texture information. In: *International Conference on Cognitive Systems and Signal Processing*, 104–121. Singapore: Springer Nature Singapore https://doi.org/10.1007/978-981-99-8018-5_8.

20 Gong, Y., Qiu, H., Liu, X. et al. (2024). Research and application of deep learning in medical image reconstruction and enhancement. *Frontiers in Computing and Intelligent Systems* 7 (3): 72–76. https://doi.org/10.54097/8w12d064.

21 Wang, W., Hu, Y., Luo, Y., and Zhang, T. (2020). Brief survey of single image super-resolution reconstruction based on deep learning approaches. *Sensing and Imaging* 21 (1): 21. https://doi.org/10.1007/s11220-020-00285-4.

22 Christopher, A., Hari Kishan, R., and Sudeep, P.V. (2022). Image reconstruction using deep learning. In: *Machine Learning Algorithms for Signal and Image Processing* (ed. D. Ghai, S.L. Tripathi, S. Saxena, et al.), 65–87. Wiley https://doi.org/10.1002/9781119861850.ch5.

23 Chen, Y., Ye, X., and Zhang, Q. (2021). Variational model-based deep neural networks for image reconstruction. In: *Handbook of Mathematical Models and Algorithms in Computer Vision and Imaging*, 1–29. Springer International Publishing https://doi.org/10.1007/978-3-030-03009-4_57-1.

24 Chen, Z., Hao, Q., and Liu, L. (2024). Research on image super resolution reconstruction based on deep learning. In Advanced Hybrid Information Processing, 435–450. https://doi.org/10.1007/978-3-031-50546-1_29.

25 Chen, Z., Pawar, K., Ekanayake, M., et al. (2022). Deep Learning for Image Enhancement and Correction in Magnetic Resonance Imaging—State-of-the-Art and Challenges. *Journal of Digital Imaging* 36 (1): 204–230. https://doi.org/10.1007/s10278-022-00721-9.

26 Zhang, H., Chen, B., Gao, X. et al. (2024). FusionOpt-Net: a transformer-based compressive sensing reconstruction algorithm. *Sensors* 24 (18): 5976. https://doi.org/10.3390/s24185976.

27 Heckel, R., Jacob, M., Chaudhari, A. et al. (2024). Deep learning for accelerated and robust MRI reconstruction. *Magnetic Resonance Materials in Physics, Biology and Medicine* 37 (3): 335–368. https://doi.org/10.1007/s10334-024-01173-8.

28 Trigka, M. and Dritsas, E. (2025). A comprehensive survey of deep learning approaches in image processing. *Sensors* 25 (2): 531. https://doi.org/10.3390/s25020531.

29 Pathrabe, T., Joy, K.K., Lal, S.S. et al. (2023). A review of image super-resolution approaches based on deep learning methods. *IET Conference Proceedings* 2023 (5): 84–91. https://doi.org/10.1049/icp.2023.1472.

30 Mastriani, M. (2014). Optimal estimation of states in quantum image processing, *arXiv preprint arXiv:1406.5121*, June.

4

Integrative Approaches in Image Analysis and Signal Interpretation

Tanishq Soni[1], Deepali Gupta[1], and Mudita Uppal[2]

[1] *Chitkara University Institute of Engineering and Technology, Chitkara University, Punjab, India*
[2] *School of Computer Science and Engineering, IILM University, Greater Noida, UP, India*

4.1 Introduction

Heart is present slightly left to center of chest which is a muscular organ. The heart's functions include pumping blood throughout the body, supplying nutrients and oxygen to tissues, and eliminating waste [1]. The heart is made up of two lower chambers called ventricles and two higher chambers called atria. The right ventricle sends oxygen-depleted blood to the lungs for oxygenation after the right atrium receives it from the body [2]. The left atrium then draws oxygenated blood into the left ventricle, which circulates it throughout the body after the lungs oxygenate it [3]. Aortic and pulmonary valves control blood flow from the ventricles into the aorta and pulmonary arteries, whereas the tricuspid and mitral valves control blood flow between the atria and ventricles [4]. Septum muscles separate the heart's left and right sides, preventing the mixing of oxygenated and deoxygenated blood. Arteries surrounding the heart provide oxygen and nourishment to the myocardium, branching into smaller arteries to reach all heart muscles [5]. Atherosclerosis, characterized by plaque build-up in the arteries, often leads to heart disease and blockages. Plaque, containing cholesterol, fat, calcium, and other substances, narrows the arteries, reducing blood flow to the heart muscle [6]. Severe blockages can result in heart attacks and other cardiovascular diseases (CVDs). The basic structure of heart is shown in Figure 4.1 which located to the left of the center of the chest and about the size of a fist [7].

The heart contracts regularly to pump blood through the circulatory system. Oxygenated blood from the lungs is received by the left heart, which then distributes it throughout the body [9]. Deoxygenated blood is pumped to the lungs by

Biomedical Imaging Technology: Signal Processing Strategies and Innovations, First Edition.
Edited by Ayush Dogra, Shalli Rani, and Ankita Sharma.

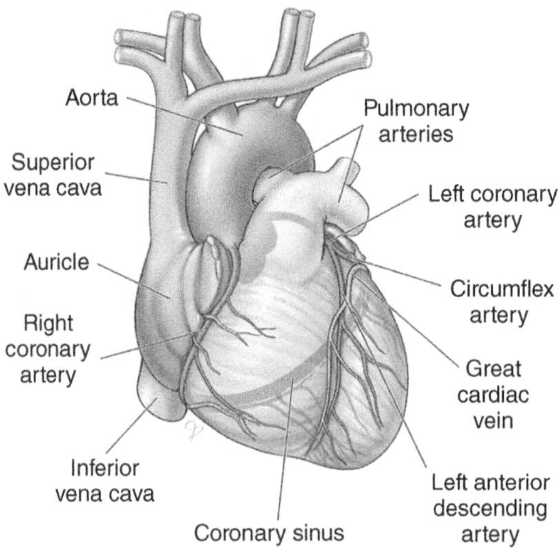

Figure 4.1 Anatomy of human heart [8]. *Source*: Cleveland Clinic.

the right heart. The upper and lower vena cava deliver blood to the heart, where the right atrium processes deoxygenated blood before pumping it into the right ventricle [10]. The pulmonary artery subsequently transports this blood to the lungs, where it releases carbon dioxide and absorbs oxygen. The pulmonary veins carry oxygen-rich blood back to the heart, where it is deposited in the left atrium before being pumped into the left ventricle [11]. From there, it is pumped out via the aorta, providing oxygenated blood to all organs, tissues, and cells, while removing CO_2 and waste. The heart regulates blood pressure by adjusting contraction force and tempo, ensuring adequate blood supply to tissues and organs [12]. Its pumping rate and strength vary according to bodily demands, increasing during exercise to supply muscles with oxygen and nutrients, and decreasing during rest to conserve energy [13]. Ultimately, life depends on the heart's ability to continually pump blood and deliver oxygen and nutrients to every muscle and organ.

There is a wide range of severity and symptoms that can be associated with cardiac disorders. Although some may not create any symptoms or may only cause slight pain, others may pose a significant risk to a person's life and have a significant influence on their quality of life [14]. The risk factors for CVD include high blood pressure, high cholesterol, diabetes, smoking, obesity, a family history of the condition, and a sedentary lifestyle. Among them are valvular heart diseases, congenital heart abnormalities, arrhythmias, heart failure, and coronary artery disease. Genetics, way of living, infections, and other medical disorders may all be the cause of these diseases. They may cause limb edema, weariness, disorientation, chest pain, and dyspnea [15]. To treat heart diseases, one often has to make lifestyle, medication, and even surgical changes. Heart disease claims the lives of over 17.9 million people a year [16]. CVDs include heart failure, heart attacks, and strokes. Heart attacks and strokes account for most CVD fatalities; they may be avoided by lowering risk factors including smoking, inadequate diet, inactivity, and alcoholism [17]. A major risk factor for stroke and heart disease is hypertension which affects one in three people over 30 globally. Further, reducing the risk of CVD is early diagnosis and treatment of diabetes and hypertension. Through balanced diets, regular exercise and medications for cardiovascular risk factors, WHO assists countries in putting CVD prevention and control plans into place. Dietary restrictions stress healthier eating and reduce salt, cholesterol, and saturated fats. Exercise may have to be cut down or limited, which calls for care. Medication management has to become second nature to control the condition [18]. For safety and well-being, social and recreational activities sometimes necessitate modifications. So, cardiac patients should adapt to their new lifestyle and have full and healthy lives with the right care and support [19]. Heart conditions can affect the heart's capacity to pump blood, raising the risk of arrhythmias, heart attacks, and heart failure. Figure 4.2 shows several common cardiac diseases.

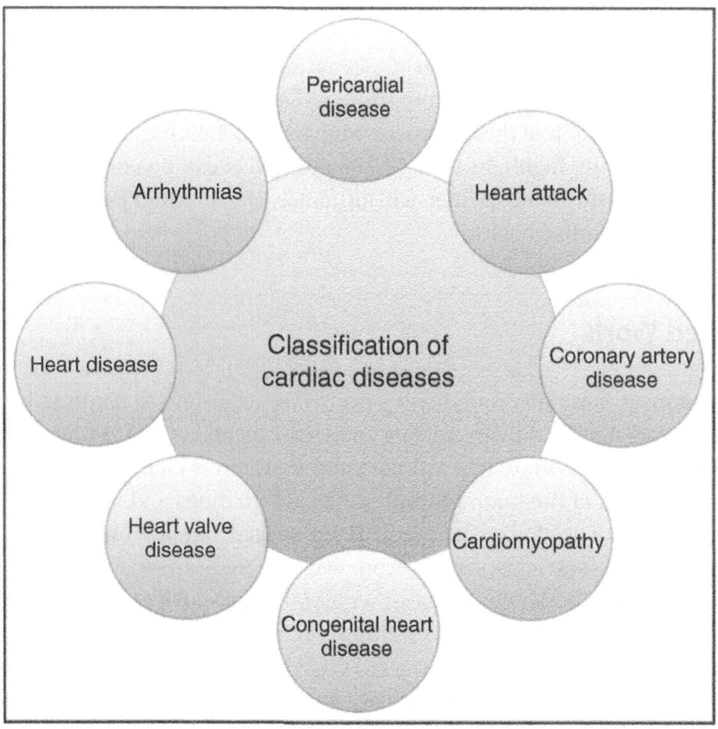

Figure 4.2 Classification of cardiac diseases.

The most common cardiac disease, coronary artery disease, occurs when plaque narrows or blocks the arteries feeding blood to the heart muscle, leading to angina, heart attacks, or myocardial infarction. Heart attacks can result from prolonged blood flow restriction to a heart region, causing the death of cardiac muscle cells and presenting symptoms such as chest discomfort, breathlessness, nausea, and light-headedness [20]. Heart failure, characterized by the heart's inability to pump enough blood, may be caused by coronary artery disease, hypertension, or cardiomyopathy, manifesting in fatigue, shortness of breath, leg swelling, and exercise difficulty. Arrhythmias, including tachycardia, bradycardia, or irregular heartbeat, are common, with symptoms such as palpitations [21]. Heart valve disease occurs when valves fail to open or close properly, disrupting blood flow and causing chest pain, tiredness, palpitations, and shortness of breath. Cardiomyopathy, resulting from improper heart muscle function due to various factors, presents symptoms such as fatigue, edema, breathlessness, and arrhythmias. Pericardial disorders, involving inflammation or anomalies in the heart's sac, may lead to conditions like pericarditis, pericardial effusion, or constrictive pericarditis [10, 22].

The synergy of cutting-edge computational tools for improved medical imaging outcomes is highlighted in this chapter, which explores integrative approaches in image analysis and signal interpretation. The study focuses on the usage of CNN and RNN in early heart disease prediction by evaluating their performance. Better diagnostic accuracy results from the more thorough signal interpretation made possible by the incorporation of these deep learning (DL) models into image analysis pipelines. The results highlight how CNNs and RNNs can be combined to maximize prediction performance, which will ultimately lead to better patient care and early detection in cardiac health.

4.2 Related Work

CVDs cause the most morbidity and death worldwide, according to Amin et al. [23]. One of the most significant clinical data analysis subjects is CVD prediction. The healthcare industry accumulates vast data and to discover characteristics to predict CVD is the need of the hour. Multiple data combinations and seven categorization methods were used to create prediction models. Examples of strategies include *k*-nearest neighbors, logistic regression, decision trees, naive Bayes, neural networks, SVM, and vote, a hybrid strategy combining naïve Bayes and logistic regression. Vote, the most effective data mining method, predicts heart disease with 87.4% accuracy.

Katarya et al. [24] surveyed that physicians have to struggle a lot to forecast and identify cardiac disease. Hospital and clinic's heart disease treatments are expensive and drinking, smoking, and inactivity cause most of the heart disease.

CVDs are predicted using supervised ML methods as ANN, DT, RF, SVM, NB, and k-nearest neighbor. The authors concluded that there is a requirement to optimize feature selection for heart disease prediction and forecast heart diseases using ML to improve outcomes.

According to Soni et al. [25], the data mining technique has been implemented in many sectors, whereas medical research uses data mining to find database information. Using basic data mining, 15 critical characteristics such as time series, clustering, association rules, and soft computing are helpful in predicting heart attack where cluster categorization fails. The small data size that finds the correct factors for heart disease prediction improves accuracy of decision tree and Bayesian classification. Naive bayes has accuracy of 96.5%, decision tree has 99.2%, and clustering has 88.3% accuracy. They concluded that a knowledge-rich environment obtained from data mining enhances treatment decisions.

Faieq et al. [26] suggested that early diagnosis is essential for predicting cardiac problems and extending life. They used UCI ML Repository which has 170 medical reports and two approaches, namely, SVM and ANN were used to diagnose heart disease. The data of 170 people were used to produce 90 training and 80 testing datasets. The accuracy and sensitivity of both approaches are compared to check which predicts each disease better. Support vector machine exhibits better accuracy of 89.1%. Subhadra et al. suggested that heart disease diagnosis is difficult in medicine. The patient's clinical tests and health history must be carefully reviewed for this diagnosis. ML aspires to construct smart automated systems to help physicians to diagnose and cure ailments. The proposed system inputs 14 clinical characteristics and trains the network using back-propagation learning. Multilayer perceptron (MLP) neural network was used to anticipate and diagnose heart disease. The recommended method diagnoses cardiac disease using 14 key factors, and the proposed MLPNN approach has 94% accuracy.

Saboor et al. [27] applied ML algorithms to detect cardiac patients and healthy persons. Hyper parameters are optimized and SVM performed best with an accuracy of 96.72%. Suresh et al. [28] proposed a hybrid random forest (RF)-support vector machine model that predicts heart diseases. Indian Pima cardiac disease dataset is selected which is available on UCI, Irvine data repository. The heart disease dataset has 13 characteristics, and SVM–RF hybrid modes predicted heart disease with 98.3% accuracy. It is more efficient and precise than comparable models. Barhoom et al. stated that early diagnosis of cardiac issues is necessary as heart disease prevalence is rising every day. The authors used RF and DL models with different classifiers where DL model peaked at 92.23% accuracy. The authors found that DL model had the greatest accuracy among other algorithms. Sharma et al. deployed deep neural network using Talos optimizer and compared it to other techniques to check which is more efficient. The selected heart disease database has 14 variables and 303 columns for patient data. Over other optimizations, Talos is more accurate having accuracy of 90.76%. Using electrocardiographic data to monitor cardiac activity, Lomoio et al. [28] found disease-related anomalies in electrocardiographic data. Autoencoders (AEs) are used to analyze signals in many fields, whereas convolutional autoencoders (CAEs) excel in signal anomaly detection. The authors trained CAE using synthetic and real 12-lead ECG signal windows. CAE has simulated and actual test set having ROC AUCs values as 97.82% and 99.75% [29]. Most of the ECG feature extraction techniques employ rule-based approaches, according to Jang et al. Manually specifying all ECG characteristics is difficult, so, CVAE was trained using 596,000 ECG samples from 1278 individuals admitted in critical care unit. Two approaches were used to verify features using three external datasets. CVAE performance was assessed by two parameters, namely, reconstruction error and Kullback divergence. Reconstruction error depends on the feature vector and the anomaly score depends on CVAE feature. About 83.5% accuracy was achieved through it [31]. Karpinski et al. described how AE neural networks can identify and fix ECG heartbeat outliers. The authors used the open-source Physio net ECG-ID database where data was collected from 90 users – 44 men and 46 women. Some users can only record a few recordings every day, so, some users have over twenty recordings collected routinely over six months. Standard deviation is used in the AE approach and highest accuracy achieved was 95.16% [32].

According to Nils Strodthof et al., cardiovascular disorders are a primary cause of death globally, making ECGs a crucial diagnostic tool because to their simplicity, non-invasiveness, and diagnostic usefulness. To efficiently analyze large volumes of time-series ECG data, automated analysis techniques, powered by machine learning and AI, have been developed. The PTB-XL+ dataset incorporates extensive ECG characteristics from commercial and open-source techniques, advancing ECG. The PTB-XL+ dataset available on Physio net includes 21,799 clinical 12-lead electrocardiograms (ECGs) from 18,869 individuals, each taken for 10 seconds. Additionally, the PTB-XL+ dataset incorporates ECG characteristics derived from the original signals using algorithms and automated diagnosis [33]. Zavadjil highlighted that the main cause of death worldwide is CVDs that include arrhythmias; thus, early and accurate diagnosis is essential for treatment. The PTB-XL+ dataset, which consists of 21,799 clinical 12-lead ECGs from 18,869 patients lasting 10 seconds each, is subjected to a decision tree model for feature extraction. Data preparation, key characteristic identification, decision tree classifier training, and computation of accuracy, precision, recall, F1-score, classification report, and confusion matrix are all part of the process. In comparison to other supervised machine learning techniques, the decision tree classifier model achieved an 84% score to guarantee optimal performance [34]. Ram et al. highlighted that the significant heart disease mortality rate requires effective and early diagnosis methods. Computer tomography, magnetic resonance imaging, impedance cardiography (ICG), ECG, and heart sound can all be used to identify cardiac illness. The ECG output was used to diagnose heart illness using three DL models: restricted Boltzmann machines (RBMs), MLPs, and Deep Belief Networks (DBNs). With accuracy of 98.6%, 97.4%, and 96.2% on the MIT-BIH dataset and 97.1%, 96.4%, and 95.3% on the PTB-ECG dataset, the proposed hybrid model performed better than existing methods [35].

4.3 Materials and Methodology

The methodology for proposing DL model that can accurately predict cardiac problems is discussed in this section. Two DL models are proposed and compared with one another. This section gives an overview of the steps that need to be taken to get the most accurate results from the proposed DL models.

4.3.1 Description of Dataset

CVD can be present in several forms which can be seen on an ECG, a diagnostic test that tracks the heart's electrical activity [30, 31]. Typical ECG results in cardiac disease consist of irregular heart rhythms, indications of heart muscle injury, issues with electrical conduction, and indications of an enlarged or strained heart. Electrocardiography is a valuable tool for diagnosing cardiac diseases, evaluating their seriousness, informing treatment choices, and tracking patients' reactions to therapy. However, accuracy is not 100% when doctors diagnose it [32]. So, DL is used to predict cardiac disease with more accuracy. Therefore, PTB-XL+ dataset having ECG signals, which is the upgraded version of PTB-XL is used in this work [33]. Each of the 21,799 clinical 12-lead ECGs in this dataset is 10 seconds long and was obtained from 18,869 patients [35]. This dataset is a long-term project being conducted at the Physikalisch-Technische Bundesanstalt (PTB), and the project's data is accessible on PhysioNet [34]. The PTB-XL+ dataset was released as an addendum to the PTB-XL dataset, comprising ECG characteristics derived from the original signals by diverse techniques [35].

There are different classes in this ECG signal dataset, namely, Normal, Myocardial Infection (MI), Conduction Disturbance (CD), ST/T Change (STTC), and Hypertrophy (HYP) as shown in Table 4.1 and Figure 4.3.

ECGs track the heart's electrical impulses to identify rhythm and function problems, helping diagnose cardiac diseases including arrhythmias and heart attacks. Figure 4.4 shows the ECG signals of different classes.

The above-mentioned figure is description of the dataset based on ECG signals.

Table 4.1 Percentage of classes in dataset.

Classes	Percentage
Normal	55.8%
Myocardial infection	15.6%
ST/T change	14.8%
Conduction disturbance	10.5%
Hypertrophy	3.3%

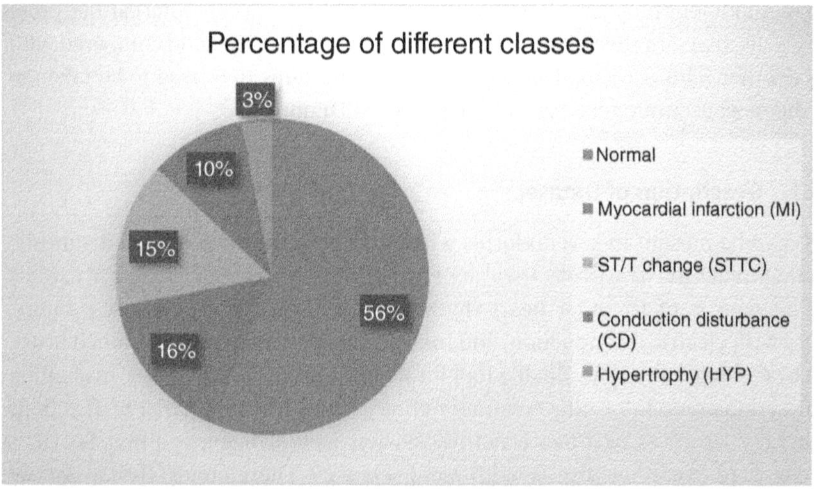

Figure 4.3 Percentage of different classes of input dataset.

4.3.2 Proposed Methodology

A research technique is necessary as it ensures that the study will be done systematically and methodically, which allows researchers to provide reliable and legitimate findings. Figure 4.5 illustrates the proposed workflow that will be used for this work.

This proposed methodology has two proposed DL models that are being evaluated.

4.3.2.1 Input Dataset and Pre-Processing

There are different sources on which dataset is available like Kaggle, UCI machine learning repository, and many more, but the PTB-XL+ dataset is collected from the PhysioNet medical dataset repository. This dataset is based on a cardiac disease

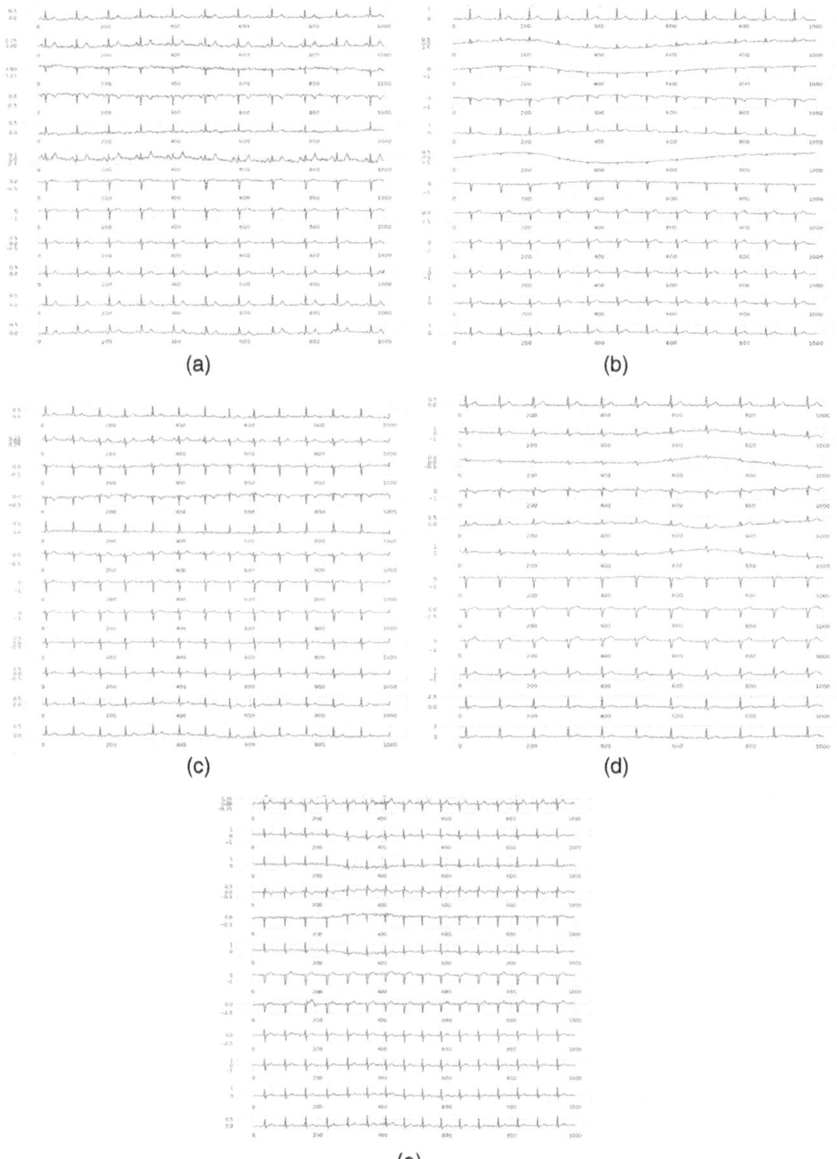

Figure 4.4 ECG signals of different classes (a) normal, (b) myocardial infection, (c) ST/T change, (d) hypertrophy, and (e) conduction disturbance.

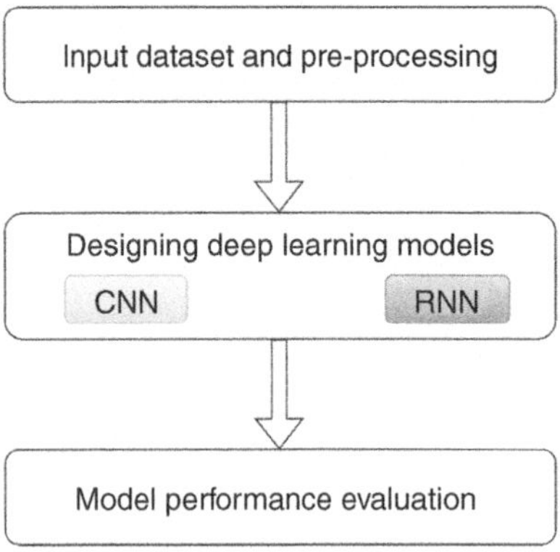

Figure 4.5 Proposed methodology.

Table 4.2 Percentage of class after data pre-processing.

Classes	Percentage
Normal	55.8%
Myocardial infection	15.6%
Other	28.6%

that has 12-lead ECG signals for 10 seconds length each. There are certain steps which are involved in data pre-processing like noise reduction, handling missing data, and feature extraction.

From the dataset, some of the features are extracted and there are five different classes in the dataset. After feature extraction, three classes are combined for better results as shown in Table 4.2 and Figure 4.6.

Following pre-processing, the data is divided into an 80%–10%–10% ratio, with 80% going toward training, 10% going toward validation, and the last 10% going toward testing.

4.3.2.2 Designing of Deep Learning Models

DL uses artificial neural networks to represent and process data to emulate the human brain's capacity to make judgments and predictions from massive volumes of complex data. It excels in image, speech, natural language processing, and pattern recognition in big datasets; therefore, this study designs two DL models: convolutional neural network and recurrent neural network (RNN). Each cardiac disease prediction model has a unique architecture and setup.

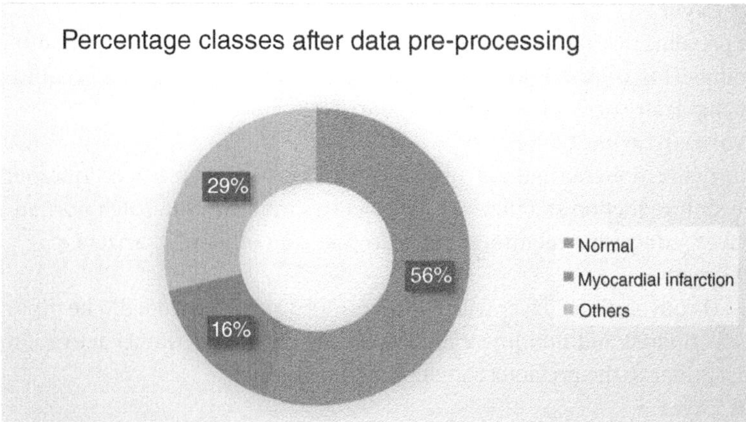

Figure 4.6 Pre-processed data proportion of classes.

4.3.2.2.1 Convolutional Neural Network In order to automatically extract patterns and features from images, CNNs use convolutional, pooling, and dense layers to learn from hierarchical representations of data. Photo-like grid-like input is ideal for CNNs. Image identification, object recognition, and picture segmentation rely on CNNs. They have the capability to learn and generalize from large picture sets. Figure 4.7 represents the proposed CNN model.

- **Conv1D Layer**
  In the input sequence, the 1D convolutional layer applies 64 kernel size 3 filters. It employs the Rectified Linear Unit (ReLU) activation function that creates non-linearity by directly outputting positive inputs and zero otherwise. The input_shape parameter defines data shape and expects input sequences with dimensions (X_train_reshaped.shape [1], 1), where X_train_reshaped.shape [1] is the number of features in each sequence.

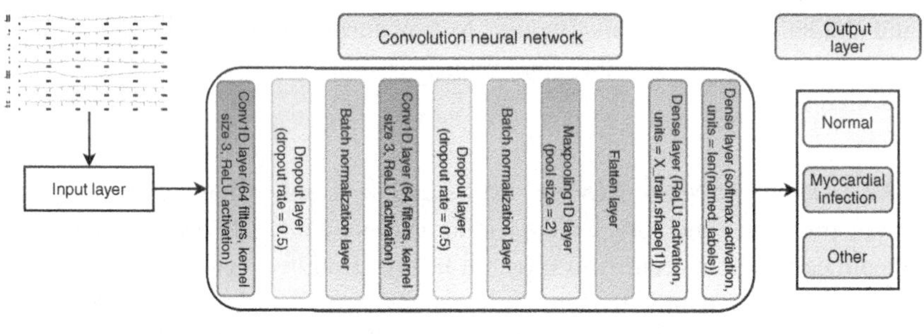

Figure 4.7 Proposed convolutional neural network.

- **Dropout Layer**
  Dropout prevents overfitting by randomly setting 50% of input units to zero during training. This layer randomly loses units, forcing the network to acquire stronger characteristics. It has a dropout rate of 0.5.
- **Batch Normalization Layer**
  Batch normalization normalizes prior layer activations per batch. Internal covariate shift reduction stabilizes and accelerates training. The batch normalization layer reduces the requirement for dropout and other regularizers.
- **Conv1D Layer**
  Another 1D convolutional layer with 64 filters and 3 kernels is added. The input sequence is filtered and non-linearity is introduced using the ReLU activation function, similar to the previous convolutional layer.
- **Dropout Layer**
  A 0.5 dropout rate dropout layer is developed for the same purpose.
- **Batch Normalization Layer**
  An additional batch normalization layer with the same features is added.
- **MaxPooling1D Layer**
  Max pooling lowers input data spatiality as it uses max pooling with a pool size of 2, halving feature map length. It keeps the most critical data while lowering computational complexity and over fitting.
- **Flatten Layer**
  This layer flattens incoming data into a one-dimensional array. It removes spatial structure to prepare data for a fully linked layer.
- **Dense Layer**
  Layer with X_train_reshaped.shape [1] neurons fully linked is added afterward. The incoming data is linearly processed before the ReLU activation function. This layer allows the network learn complicated data patterns.
- **Dense Layer**
  The last completely connected layer with as many neurons as classes (assuming named_labels includes class labels) is added. It uses a linear operation and the softmax activation function to transform the raw output into class probability scores where the maximum probability class is projected as the output class.

4.3.2.2.2 Recurrent Neural Network RNNs are designed to handle consecutive input by preserving memory. Unlike feedforward networks, RNNs include recurrent connections that retain and transmit information between sequence stages. RNNs can analyze data with temporal dependencies of various durations since they can handle input sequences of different lengths. However, the vanishing gradient issue limits traditional RNNs' ability to capture distant interactions. Researchers developed LSTM and GRU models to solve this issue. These models

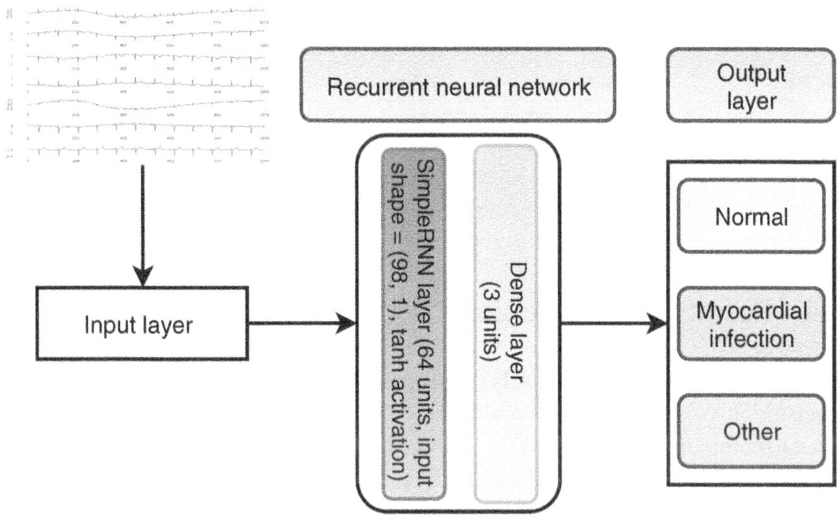

Figure 4.8 Recurrent neural network.

use specialized methods to preserve and update knowledge, improving long-term relationship learning. Figure 4.8 shows the proposed RNN layers.

- **Sequential Model**
  Sequential model layers are linear that enables to create a neural network by adding layers sequentially.
- **Simple RNN Layer**
  This is a simple RNN layer that sets the output space dimensions and layer neurons. In this model, it is set to 64 units. This option determines the layer's expected input data shape. This sequence has 98 time steps and 1 feature and layer's unit output activation function. The hyperbolic tangent function (tanh) limits output values to $[-1, 1]$.
- **Dense Layer**
  Dense layers have completely linked neurons that connect current layer and the previous layer. It sets output space dimensions and layer neuron count. In this, it is set to 3 units. The default activation function is linear.
- **Output Layer**
  Using the output produced by the recurrent layer, the output layer is in charge of creating the network's final output. The output layer is composed of a single node or numerous nodes, depending on the activity that is being performed. In this, the result is classified into 3 classes, namely normal, MI and others.

These mentioned steps are being followed for the research in the research methodology. Two different DL models are being proposed and in next section, these will be evaluated on various parameters.

Table 4.3 Simulation parameters.

Sr. No.	Parameters	Values
1	Batch size	32
2	Optimizer	Adam
3	Epochs	50

4.4 Results and Discussion

For PTB-XL+ analysis, two models are proposed which are examined on accuracy, precision, loss, F1-Score, and recall. The PTB-XL+ dataset was thoroughly analyzed to compare CNNs and RNNs. CNNs captured spatial hierarchies in sequential data and RNNs represented temporal relationships. This work illuminates the best ways to handle complicated and large real-world datasets. Table 4.3 shows the simulation parameters used for training and testing DL models.

4.4.1 Analysis Based on Confusion Matrix

Confusion accuracy, also known as confusion matrix accuracy or classification accuracy, measures classification model performance. Figure 4.9 shows and compares the confusion matrix of 2 DL models.

A number of significant metrics that characterize a classification model's performance are computed using the confusion matrix. Accuracy, precision, recall, loss, and F1-score are computed from this.

4.4.2 Analysis Based on Accuracy

In classification applications, DL model validation and training accuracy are crucial. When the model memorizes training data without learning patterns, high training accuracy and low validation accuracy may indicate overfitting. Figure 4.10 compares validation and training accuracies of 3 DL models.

Accuracy of CNN has been reached at 80%, whereas the accuracy of RNN is 72%.

4.4.3 Analysis Based on Loss

DL, especially neural network training, requires training and validation loss. Training loss is minimized using gradient descent optimization. Validation loss detects over fitting, a situation in which the model over adapts to training data and fails to generalize. Both losses must decrease for effective learning without over fitting. The analysis of training and validation loss is presented in Figure 4.11.

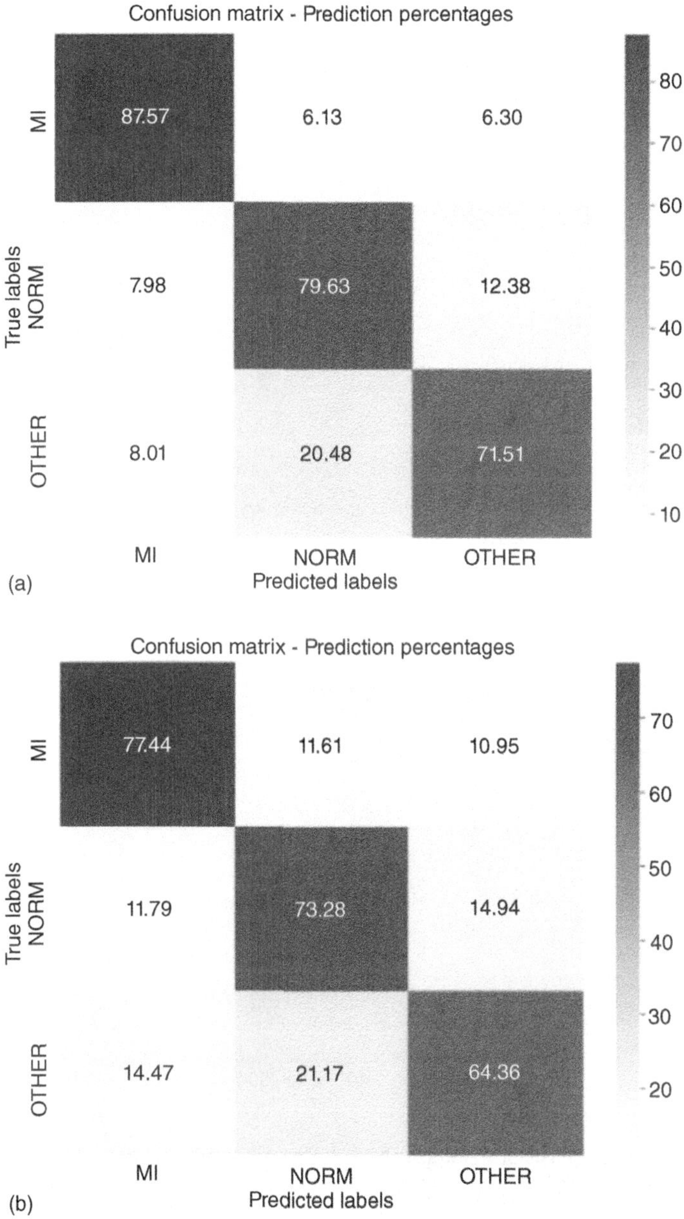

Figure 4.9 Comparison on the basis of confusion matrix (a) CNN and (b) RNN.

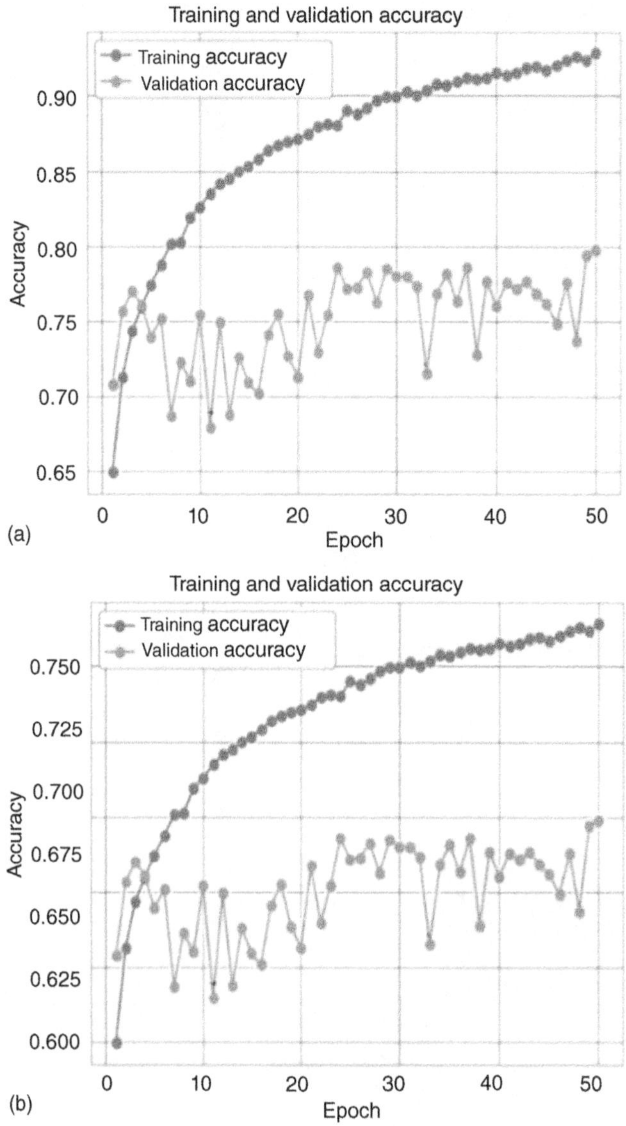

Figure 4.10 Comparison on basis of training and validation accuracy (a) CNN and (b) RNN.

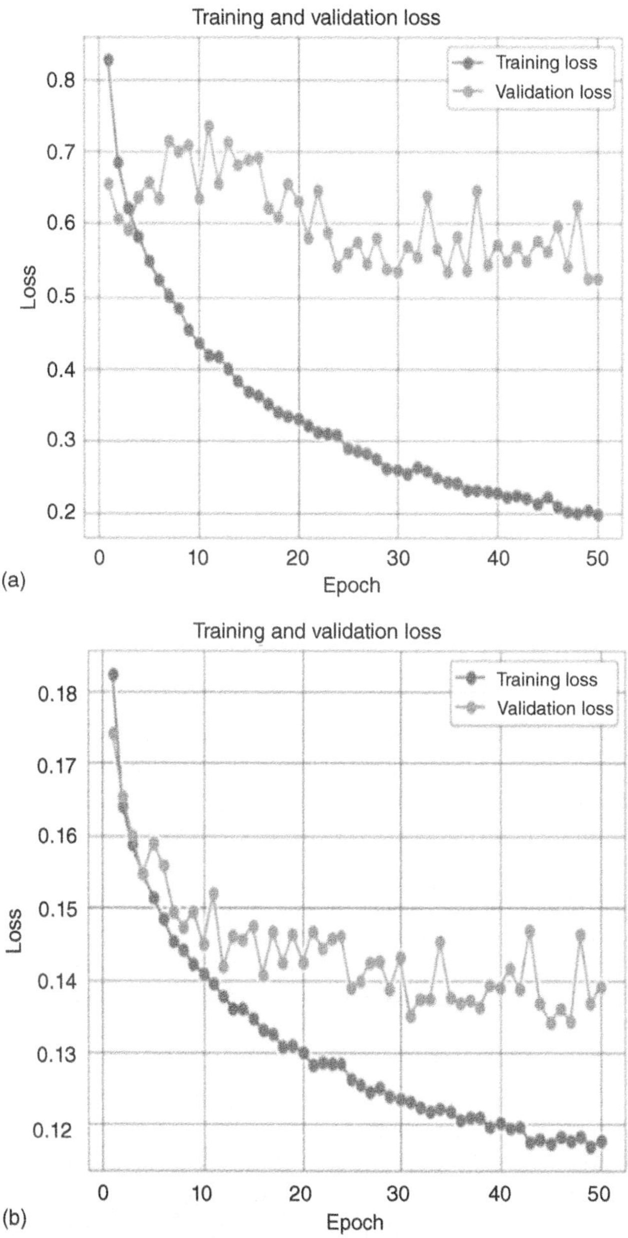

(a)

(b)

Figure 4.11 Comparison on basis of training and validation loss (a) CNN and (b) RNN.

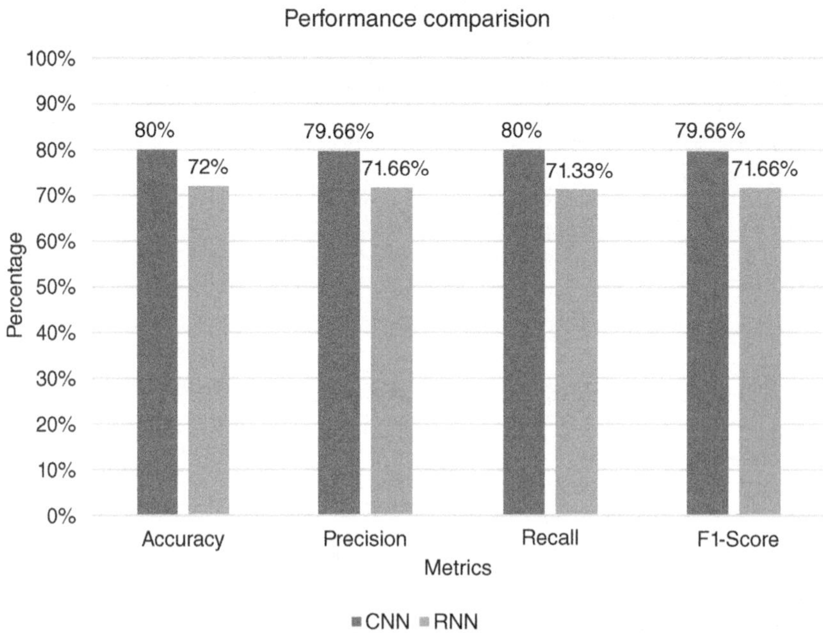

Figure 4.12 Performance comparison of proposed model.

These models are evaluated using 32 batch size, an Adam optimizer, and 50 epochs. These two models' accuracy, f1-score, precision, and recall may be calculated using confusion matrix parameters. This study shows that CNNs outperform alternative designs on the PTB-XL+ dataset. CNNs have the best accuracy as shown in Figure 4.12.

With the cardiac disease prediction dataset PTB-XL+, two models were proposed and tested. This comparison compares accuracy, f1-score, recall and precision. The study concluded that CNNs were the best model architecture. CNNs outperformed RNNs in recognizing subtle dataset patterns. This extensive examination shows that CNNs are adaptable and superior at handling large-scale sequential data applications like language modeling.

The findings show that the prediction capacities for early heart disease identification are much improved by image analysis with CNNs and RNNs. These methods improve diagnosis accuracy and reliability by utilizing the advantages of RNNs in sequential data processing and CNNs in spatial feature extraction. These findings demonstrate how important it is to use a DL model in order to optimize medical imaging and signal interpretation, which will ultimately help develop more successful early intervention techniques for heart disorders.

4.5 Conclusion and Future Scope

Cardiac disease detection is important as it is the main cause of death worldwide. Early cardiac disease detection allows prompt therapies that enhance patient outcomes and prevent complications. ECG is essential for cardiac disease identification, treatment, and prevention, lowering medical expenses and enhancing patient outcomes. The PTB-XL+ dataset, which is based on ECG signals, is taken to identify cardiac disease at an extremely early stage. DL algorithms applied to the given dataset provide advances in the capacity to forecast cardiac diseases. A total of 21,799 clinical 12-lead ECGs taken from 18,869 patients lasts for 10 seconds. The study shows that integrating CNN and RNN models with advanced image processing techniques improves the accuracy and reliability of early cardiac disease prediction, opening the door to better diagnostic results. For testing, these two are being subjected to 50 epochs, Adam optimizer, and 32 batch sizes. A comparison is done on the basis of accuracy, F1-score, precision, and recall. Between these two neural networks, CNN has achieved the maximum accuracy of 80%, while RNN has achieved an accuracy of 72%. CNN has a precision value of 79.66%, and RNN has a precision of 71.66%. This shows that CNN model outperforms than RNN.

There are a variety of deep-learning models that could be utilized in future studies to make predictions regarding cardiac disease. To achieve better results, it is possible to expand the size of the dataset. In order to obtain better results, increase the number of epochs, batch size or switching of optimizers can be done. Also, by constructing a hybrid model on the ECG signals dataset, it would be possible to attain the higher accuracy in the future.

References

1 Nandy, S., Adhikari, M., Balasubramanian, V. et al. (2023). An intelligent heart disease prediction system based on swarm-artificial neural network. *Neural Computing and Applications* 35 (20): 14723–14737.

2 Yaqoob, M.M., Nazir, M., Khan, M.A. et al. (2023). Hybrid classifier-based federated learning in health service providers for cardiovascular disease prediction. *Applied Sciences* 13 (3): 1911.

3 Shukur, B.S. and Mijwil, M.M. (2023). Involving machine learning techniques in heart disease diagnosis: a performance analysis. *International Journal of Electrical and Computer Engineering* 13 (2): 2177.

4 Das, R.C., Das, M.C., Hossain, M.A., et al. (2023, March). Heart disease detection using ML. In *2023 IEEE 13th Annual Computing and Communication Workshop and Conference (CCWC)* (pp. 0983–0987). IEEE.

5 Miah, J., Ca, D.M., Sayed, M.A., et al. (2023, November). Improving cardiovascular disease prediction through comparative analysis of machine learning models: a case study on myocardial infarction. In *2023 15th International Conference on Innovations in Information Technology (IIT)* (pp. 49–54). IEEE.

6 Oluwasanmi, A., Aftab, M.U., Baagyere, E. et al. (2021). Attention autoencoder for generative latent representational learning in anomaly detection. *Sensors (Basel)* 22 (1): 123.

7 Shan, L., Li, Y., Jiang, H. et al. (2022). Abnormal ECG detection based on an adversarial autoencoder. *Frontiers in Physiology* 13: 961724.

8 Cleveland Clinic. Heart: Anatomy and function [Internet]. Cleveland Clinic. https://my.clevelandclinic.org/health/body/21704-heart (accessed 17 Sept 2024).

9 Roy, T.S., Roy, J.K., and Mandal, N. (2022). Classifier identification using deep learning and machine learning algorithms for the detection of valvular heart diseases. *Biomedical Engineering Advances* 3: 100035.

10 Nancy, A.A., Ravindran, D., Raj Vincent, P.D. et al. (2022). IoT-cloud-based smart healthcare monitoring system for heart disease prediction via deep learning. *Electronics* 11 (15): 2292.

11 Tran, B.X., Latkin, C.A., Vu, G.T. et al. (2019). The current research landscape of the application of artificial intelligence in managing cerebrovascular and heart diseases: a bibliometric and content analysis. *International Journal of Environmental Research and Public Health* 16 (15): 2699.

12 Jiang, F., Jiang, Y., Zhi, H. et al. (2017). Artificial intelligence in healthcare: past, present and future. *Stroke and Vascular Neurology* 2 (4): 230–243.

13 Sk, K.B., Roja, D., Priya, S.S., et al. (2023, March). Coronary heart disease prediction and classification using hybrid machine learning algorithms. In *2023 International Conference on Innovative Data Communication Technologies and Application (ICIDCA)* (pp. 1–7). IEEE.

14 Padmaja, B., Srinidhi, C., Sindhu, K. et al. (2021). Early and accurate prediction of heart disease using machine learning model. *Turkish Journal of Computer and Mathematics Education (TURCOMAT)* 12 (6): 4516–4528.

15 Kavitha, M., Gnaneswar, G., Dinesh, R., et al. (2021, January). Heart disease prediction using hybrid machine learning model. In *2021 6th International Conference on Inventive Computation Technologies (ICICT)* (pp. 1329–1333). IEEE.

16 Yu, D., Yang, S., Wang, R., et al. (2023). Machine learning in heart failure research: a bibliometric analysis from 2003 to 2023.

17 Soni, T., Gupta, D., and Uppal, M., (2023, December). Transforming the prediction of heart disease: an empirical analysis of machine learning classifiers. In *2023 IEEE Pune Section International Conference (PuneCon)* (pp. 1–5). IEEE.

18 Aggarwal, S., Gupta, S., Kannan, R. et al. (2022). A convolutional neural network-based framework for classification of protein localization using confocal microscopy images. *IEEE Access* 10: 83591–83611.

19 Kapoor, D., Gupta, D., and Uppal, M. (2024, August). A scientometric review of machine learning and deep learning techniques. In *2024 7th International Conference on Circuit Power and Computing Technologies (ICCPCT)* (Vol. 1, pp. 1465–1470). IEEE.

20 Soni, T., Uppal, M., Gupta, D., and Gupta, G. (2023, May). Efficient machine learning model for cardiac disease prediction. In *2023 2nd International Conference on Vision Towards Emerging Trends in Communication and Networking Technologies (ViTECoN)* (pp. 1–5). IEEE.

21 Soni, T., Gupta, D., Uppal, M., and Kumari, A. (2024, July). Machine learning in cardiovascular disease: clinical applications and relevance to cardiac imaging. *In 2024 2nd International Conference on Sustainable Computing and Smart Systems (ICSCSS)* (pp. 940–944). IEEE.

22 Sarmah, S.S. (2020). An efficient IoT-based patient monitoring and heart disease prediction system using deep learning modified neural network. *IEEE Access* 8: 135784–135797.

23 Amin, M.S., Chiam, Y.K., and Varathan, K.D. (2019). Identification of significant features and data mining techniques in predicting heart disease. *Telematics and Informatics* 36: 82–93.

24 Katarya, R. and Srinivas, P. (2020, July). Predicting heart disease at early stages using machine learning: a survey. In *2020 International Conference on Electronics and Sustainable Communication Systems (ICESC)* (pp. 302–305). IEEE.

25 Soni, J., Ansari, U., Sharma, D., and Soni, S. (2011). Predictive data mining for medical diagnosis: an overview of heart disease prediction. *International Journal of Computer Applications* 17 (8): 43–48.

26 Faieq, A.K. and Mijwil, M.M. (2022). Prediction of heart diseases utilising support vector machine and artificial neural network. *Indonesian Journal of Electrical Engineering and Computer Science* 26 (1): 374–380.

27 Subhadra, K. and Vikas, B. (2019). Neural network based intelligent system for predicting heart disease. *International Journal of Innovative Technology and Exploring Engineering* 8 (5): 484–487.

28 Saboor, A., Usman, M., Ali, S. et al. (2022). A method for improving prediction of human heart disease using machine learning algorithms. *Mobile Information Systems* 2022: 1–9.

29 Suresh, T., Assegie, T.A., Rajkumar, S., and Kumar, N.K. (2022). A hybrid approach to medical decision-making: diagnosis of heart disease with machine-learning model. *International Journal of Electrical and Computer Engineering (IJECE)* 12 (2): 1831–1838.

30 Barhoom, A.M., Almasri, A., Abu-Nasser, B.S., and Abu-Naser, S.S. (2022). Prediction of heart disease using a collection of machine and deep learning algorithms.

31 Sharma, S. and Parmar, M. (2020). Heart diseases prediction using deep learning neural network model. *International Journal of Innovative Technology and Exploring Engineering (IJITEE)* 9 (3): 2244–2248.

32 Lomoio, U., Vizza, P., Giancotti, R. et al. (2023). AUTAN-ECG: an autoencoder based system for anomaly detection in ECG signals. *Authorea Preprints.*

33 Strodthoff, N., Mehari, T., Nagel, C. et al. (2023). PTB-XL+, a comprehensive electrocardiographic feature dataset. *Scientific Data* 10 (1): 279.

34 Zavadjil, M. (2024). A decision tree classifier algorithm for the automated analysis of electrocardiogram signals to identify heart arrythmias. *Journal of High School Science* 8 (1): 74–86.

35 Ram, R.S., Akilandeswari, J., and Kumar, M.V. (2023). HybDeepNet: a hybrid deep learning model for detecting cardiac arrhythmia from ECG signals. *Information Technology and Control* 52 (2): 433–444.

5

Multimodal Imaging: Combining Molecular and Optical Approaches

Haewon Byeon[1], Azzah AlGhamdi[2], Ismail Keshta[3], Mukesh Soni[4], Mohammad Shabaz[5], and Mohammed Wasim Bhatt[6]

[1]*Workcare D-Health Lab, Convergence Department, Korea University of Technology and Education, Cheonan, South Korea*
[2]*Computer Information Systems Department, College of Computer Science and Information Technology, Imam Abdalrhman Bin Faisal University, Khobar, Saudi Arabia*
[3]*Computer Science and Information Systems Department, College of Applied Sciences, AlMaarefa University, Riyadh, Saudi Arabia*
[4]*Centre for Research Impact & Outcome, Chitkara University Institute of Engineering and Technology, Chitkara University, Rajpura, Punjab, India*
[5]*Department of Computer Engineering, Faculty of Engineering and Technology, Marwadi University Research Center, Marwadi University, Rajkot, India*
[6]*Model Institute of Engineering and Technology, Jammu, J&K, India*

5.1 Introduction

According to the latest statistics on female breast cancer from the American Cancer Society in 2022, the incidence of breast cancer has been on the rise since 1982. The incidence has increased by about 0.5% each year in the most recent data year [1]. The treatment approach relies on early identification and prompt action because an excellent prognosis is associated with an early breast cancer diagnosis. Currently, the Nottingham histological grading method is utilized worldwide to assess breast cancer grading [2]. Atomic atypia, mitotic count, and tubule development are this system's three primary morphological features. Among these, the mitotic count is a crucial biomarker for identifying breast cancer malignancy and is of great significance in diagnosis, treatment, and prognosis.

Early mitotic count detection methods were mainly based on traditional deep learning algorithms [3]. The performance of such algorithms depends on manual feature extraction of images, including morphological features, texture features, and statistical features, which requires pathologists to use professional knowledge to define and describe. The area of computer vision has seen tremendous strides since the advent of deep learning [4]. Using convolutional neural networks

Biomedical Imaging Technology: Signal Processing Strategies and Innovations, First Edition.
Edited by Ayush Dogra, Shalli Rani, and Ankita Sharma.

(CNNs), deep convolutional networks (DNNs) can automatically detect characteristics of mitotic cells, which is a great advantage. The IDSIA model [5] uses a deep neural network classifier to differentiate between normal cells and cells undergoing mitosis. Applying the trained classifier to images of breast tissue pathology images using a sliding window approach is done in the detection stage. Unfortunately, its inefficiency and excessive computational complexity makes it unsuitable for healthcare settings. The HC+CNN model [6] uses manual and convolutional features to design a cascade classifier. Compared with IDSIA, their computational complexity is significantly reduced, but because the selection of candidate blocks is still based on traditional cell segmentation methods, it is easy to cause the omission of mitotic cells.

Two-stage deep detection models are heavily used to detect mitosis. Two stages are involved: a coarse retrieval network finds potential blocks of mitotic cells, and a classification network sorts them. Consider the CasNN model [7], a deep cascade detection system that combines two CNNs. One uses a coarse retrieval network to identify potential blocks of mitotic cells and to rank those blocks; the other uses a classification network. The two networks are trained separately, which could lead to a less user-friendly detection system. This is one possible downside of the approach. The model DRN [8] uses a deep regression network trained on FCN to detect mitosis. The location of the mitotic centroid is deduced as the local maxima during the detection stage. Datasets lacking pixel-level annotations are inapplicable, even though this technique produces beneficial performance outcomes on fully annotated ICPR 2012 mitotic datasets. With the use of a general object identification network, the model Deep Mitosis [9] suggests a system for identifying mitotic cells in sections stained with H&E. It has similar disadvantages to DRN: for images like for mitosis datasets such as ICPR 2014, which only provide centroid annotations, the overall work efficiency is reduced due to weak supervision capabilities.

However, the existing mitosis detection methods still cannot effectively solve the problem of significant morphological changes within mitotic cells. Figure 5.1 shows a great deal of diversity in the morphological changes of chromosomes throughout the four phases of mitosis. These four phases are:

Prophase: Chromosomes condense and become visible. The nuclear envelope breaks down, and the mitotic spindle forms from centrosomes at opposite poles.

Metaphase: Chromosomes align along the metaphase plate in the center of the cell. Spindle fibers attach to the centromeres of each chromosome.

Anaphase: Sister chromatids separate as spindle fibers pull them toward opposite poles of the cell.

Telophase: Chromosomes decondense, the nuclear envelope reforms around each chromosome set, and the spindle apparatus disassembles, resulting in two new nuclei.

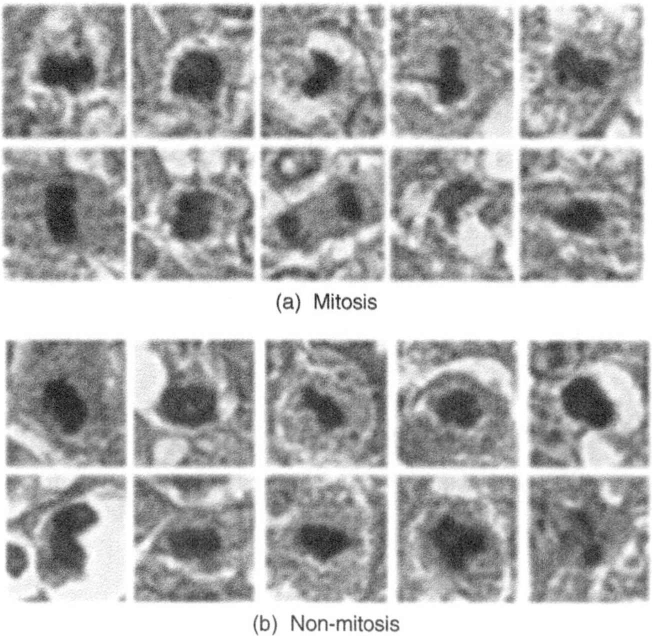

Figure 5.1 Cells that mitosis (a) and those that do not (b).

As a result, the detection results will have many false negatives (FN) and false positives (FP), which will impact the overall Model's performance and generalizability.

Mitosis recognition in breast pathology images can be challenging; this research presents a task-guided radial basis network solution to address these issues. The radial basis function (RBF) center is defined in combination with the mitotic cell identification task while detecting mitotic cells in two stages. Determining the RBF center, an important parameter, is the key to the classification problem. Traditional methods usually use unsupervised clustering algorithms to define RBF centers based on data distribution. At the same time, RBF networks are essential for processing object classification, but the overarching purpose of feature space clustering analysis must fully align with these needs. So, the RBF center that the primary clustering method produces might be challenging for the classification job. To begin, the mitotic sample features are used to initialize the RBF center by using the usual *K*-means clustering algorithm, and then the optimal radial basis center is determined by iterative update based on the classification task to use different RBF function centers to better handle the variability of mitotic cell morphological structure and enhance the precision of mitotic cell detection. Through experiments, the ICPR 2014 and AMIDA 2013 based on centroid annotation are used to identify the mitotic cells. The public dataset of mitotic cells verifies our method is advanced and superior.

Section 5.2 describes the network architecture, components, training process, and hyperparameters used, and Section 5.3 performs an assessment using various metrics, experimental setup, and comparative analysis results. Section 5.4 concludes with findings, implications, limitations, and suggestions for future research directions.

5.2 Network Model

The overall solution of the task-guided radial basis network for the breast pathology images mitosis detection model is shown in Figure 5.2, divided into three steps.

1) Collect images of the training dataset, collect positive samples containing image category information and center position information and negative samples containing only category information for the first stage network training
2) Locate mitotic cells and extract candidate objects based on the deep supervision mechanism network model
3) Input the extracted candidate blocks into the task-guided RBF network to realize the final imaging of breast cancer cells during mitosis.

Numerous critical processes are involved in creating a task-Guided Radial Basis Network (RBFN) for mitotic recognition in breast pathology images. Below is an overview of the generic algorithm that can identify mitosis in breast pathology photos. Remember that this is only a high-level overview and that successful implementation will necessitate a solid grasp of image processing and machine learning methods. The key components involved in developing the Model are as follows:

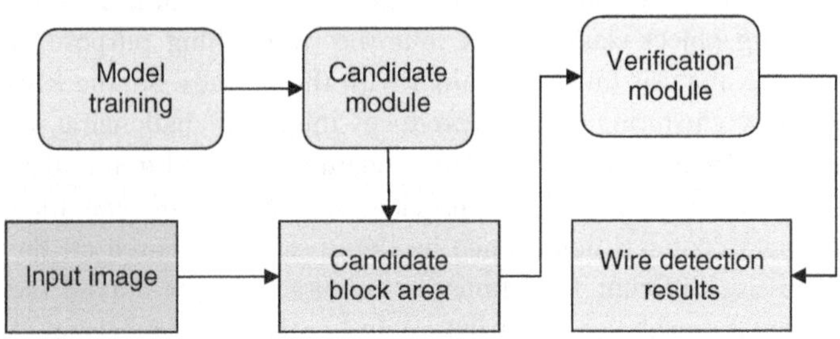

Figure 5.2 Overall frame structure.

1) Images and related annotations are loaded from the dataset using the loadDataset() function.
2) NormizeImages(images): This function adjusts the pixel values of an image.
3) The ResizeImages function resizes images to the specified scale (images, target_size).
4) Applying augmentation techniques to pictures and annotations is known as data augmentation (images, annotations).
5) ExtractFeatures(pictures): This function uses a CNN or other techniques to extract image features.
6) Initialize an RBFN with the given number of input features and RBF neurons by calling InitializeRBFN(num_features, num_rbf_neurons).
7) InitializeRBFCenters(features, num_rbf_neurons): This function uses clustering techniques such as *k*-means to initialize RBF centers.
8) InitializeRBFSpreads(features, centers): Determines the starting points for RBF spreads by measuring the separations between centers.
9) RBF layer outputs (features, centers, and spreads) are computed using the function computeRBFOutputs.
10) ComputeOutputs(rbf_outputs): Returns the network's final outputs.
11) CrossEntropyLoss(): This function defines the classification loss function.
12) SGDOptimizer(learning_rate): Specifies the training optimizer.
13) ComputeGradients(loss, RBFN): Performs backpropagation gradient computations.
14) ApplyThreshold(predictions, threshold): Determines the final classification by applying a threshold to the projections.
15) DeployModel(RBFN, deployment_environment): This function makes the trained Model available for real-world or clinical use.
16) MonitorModelPerformance(RBFN, deployment_environment): Tracks the deployed Model's performance and makes required adjustments.

The primary procedures for creating, refining, and testing an RBFN for mitosis detection are covered in Algorithm 5.1.

Algorithm 5.1 Breast pathology image classification using RBFN

1. Load and Preprocess Data
   INPUT: Dataset of breast pathology images with mitosis annotations

```
// Load images and annotations
images, annotations = LoadDataset()

// Preprocess images
images = NormalizeImages(images)
images = ResizeImages(images, target_size = (256, 256))
```

```
// Augment data
images, annotations = DataAugmentation(images, annotations)

// Extract features (e.g., using a pre-trained CNN for feature extraction)
features = ExtractFeatures(images)

2. Initialize RBFN
 INPUT: Number of features = num_features

 // Define network architecture
 RBFN = InitializeRBFN(num_features, num_rbf_neurons)

 // Initialize RBF centers and spreads
 centers = InitializeRBFCenters(features, num_rbf_neurons)
         // e.g., using k-means clustering
 spreads = InitializeRBFSpreads(features, centers)
         // e.g., based on distance between centers

3. Train RBFN
 INPUT: Training features, training labels
 OUTPUT: Trained RBFN model

 // Define loss function and optimizer
 loss_function = CrossEntropyLoss()
 optimizer = SGDOptimizer(learning_rate = 0.01)

 // Training loop
 for epoch in 1 to num_epochs do
    for batch in CreateBatches(training_features, training_labels) do
       // Forward pass
       batch_features, batch_labels = batch
       rbf_outputs = ComputeRBFOutputs(batch_features, centers, spreads)
       predictions = ComputeOutputs(rbf_outputs)

       // Compute loss
       loss = loss_function(predictions, batch_labels)

       // Backward pass and optimization
       gradients = ComputeGradients(loss, RBFN)
       optimizer.UpdateParameters(RBFN, gradients)
```

```
        // Log training progress
        LogProgress(epoch, batch, loss)

4. Evaluate Model
   INPUT: Test features, test labels
   OUTPUT: Evaluation metrics

   // Forward pass on test data
   rbf_outputs_test = ComputeRBFOutputs(test_features, centers, spreads)
   predictions_test = ComputeOutputs(rbf_outputs_test)

   // Calculate performance metrics
   accuracy  = ComputeAccuracy(predictions_test, test_labels)
   precision = ComputePrecision(predictions_test, test_labels)
   recall    = ComputeRecall(predictions_test, test_labels)
   f1_score  = ComputeF1Score(precision, recall)

   // Log evaluation results
   LogEvaluationResults(accuracy, precision, recall, f1_score)

5. Post-Processing and Deployment
   INPUT: Trained RBFN model

   // Apply post-processing techniques
   processed_results = ApplyThreshold(predictions_test, threshold = 0.5)

   // Deploy Model
   DeployModel(RBFN, deployment_environment)

   // Monitor and update
   MonitorModelPerformance(RBFN, deployment_environment)
   UpdateModelIfNecessary()
```

5.2.1 Dataset Selection

ICPR 2014 dataset [10]: The MITOSIS & ATYPIA competition released a dataset to assess nuclear atypia scoring and image analysis to detect mitosis in breast cancer histology slides using ICPR 2014 methodology. One Aperio Scanscope XT and one Hamamatsu Nanozoomer 2.0-HT slide scanner, with Ï20 and ×40 magnifications, were used to scan the slides after staining them with standard H&E dye. Typically, we employ a magnification of Ï40 to identify mitotic cells. The 1696 pathologic images that comprise the collection have a pixel size of 1539 × 1376.

The training data is 1200 images from 16 different biopsies, and the test data is 496 images from 5 different breast biopsies. In our experiments, due to the ICPR, The organizers of the 2014 competition only provided the mitosis labels on the training set while ignoring the test set labels. Thus, we used the dataset to randomly split the Training set into four parts: training data and validation data partitioning approach outlined in [11] ratio to obtain a fair experimental comparison.

AMIDA 2013 dataset [12]: Did MICCAI hold an open competition in 2013 for testing algorithms for mitosis identification? Images in this collection are derived from pathological sections stained with H&E, which were taken from 23 patients diagnosed with invasive breast cancer. Twelve patients' 311 HPF pictures, with 550 mitotic cell annotations, make up the training set. Eleven patients' 295 HPF pictures, with 533 mitotic cell annotations, make up the test set. The dimensions of every HPF image are 2000 × 2000 pixels. Multiple pathologists have annotated the mitotic cells in the dataset in a format similar to the cell centroid pixel annotation of ICPR 2014.

5.2.2 Image Patches for Classification and Regression Localization

Unlike natural images, where bounding boxes containing size and location information detect targets, pathologists manually annotate point sets on medical pathology images that do not include cell location and size information. Through some studies related to breast tissue pathology image detection, we found that most researchers choose to use image patches of fixed size as training datasets, such as stacked sparse autoencoders (SSAE) [13], separated channel CNN (SC-CNN) [14], and semi-supervised adversarial neural networks SGAN [15]. These methods collect training patches based on the image annotation information to obtain many positive and negative samples, but the collected data does not contain positioning estimation information.

In traditional image patch-based network models, when it comes to training network models, the sampled picture patches are limited to providing category labels; they cannot offer location regression data [16]. In this chapter, we propose to obtain high-quality mitotic cell candidate patches by training a new image patch-based deep learning method. In addition to distinguishing mitotic and non-mitotic cells from the background, this method can also distinguish mitotic cells from non-mitotic cells. The center position expression of the block is also added to the image block information so that the mitotic cell categorization and location analysis of mitotic cells can be processed simultaneously.

To enhance the network's classification deep learning and location ability based on image blocks, the size of the sampled image blocks is set to 60 × 60 pixels, and the positive–negative sample ratio is 1:2. The positive sample sampling is based on the manually annotated position provided in the database as the reference point. A small range of random offsets is performed (the offset in this chapter is selected as 0–10 pixels). After that, the image block is collected with the offset position as the center, and the offset is saved for the location regression task. The negative sample is generated by random sampling, and it must be ensured that its center is greater than 60 pixels from the nearest manual label (positive sample). The negative sample only contains the category information of the image block and does not participate in the location regression. Figure 5.3 shows some example image blocks used for Training, where the green "+" represents the sampled picture block's center. In contrast, the yellow "+" stands for the manually annotated kernel's center, also known as ground truth.

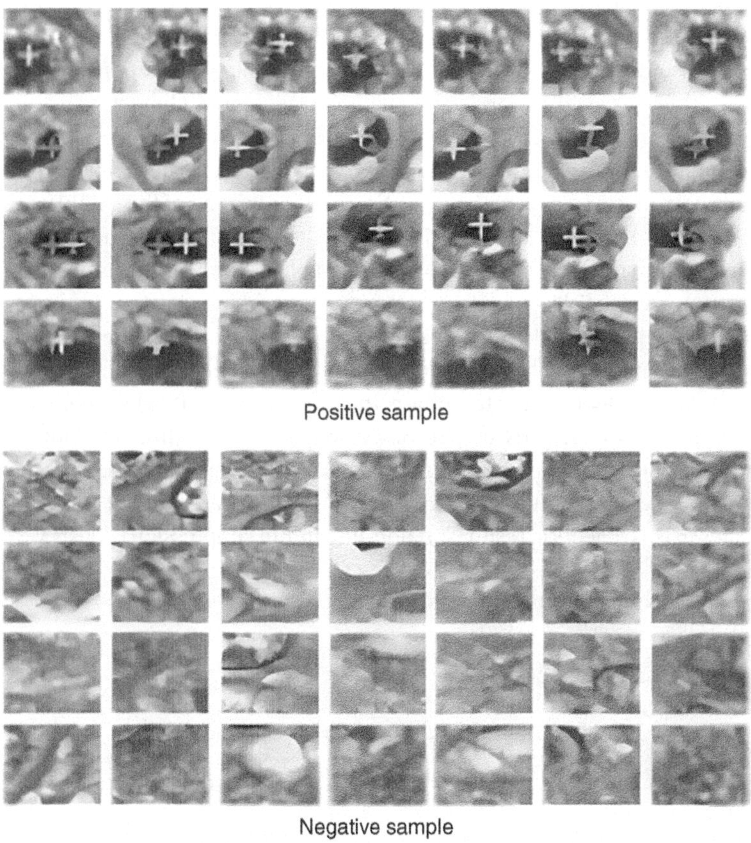

Figure 5.3 Example image patches used for network training.

5.2.3 Candidate Block Screening Network

To obtain candidate blocks of mitotic cells, this chapter constructs a fully convolutional network model that integrates deep supervision [17] and attention mechanism to screen candidate blocks of mitosis in the first stage, utilizing all the scale levels' prediction capabilities, thereby comprehensively improving the detection performance and obtaining high-quality candidates of mitosis. Figure 5.4 shows the exact network architecture.

This architecture employs a network of convolutional neural units as its foundation for feature extraction, in which 3×3 convolution kernels are mainly used, and more nonlinear information is learned through multiple convolution operations [18]. To make the more profound network architecture not use fully connected hidden layers, following each network layer, batch normalization layers, and ReLU activation functions are employed, and maximum to aggregate the features that have been retrieved, pooling layers are placed between convolutional layers [19]. All the side output layers, as well as all of the pooling layers and the last convolutional layer, use deep learning. A 1×1 convolutional block attention module (CBAM) is initially used to calibrate the channel and spatial characteristics in every side output layer path [20]. Feature dimensionality reduction and channel feature combination are accomplished with the convolution kernel's help. After that, the fully convolutional layer is used as the classifier instead of the completely connected layer. The softmax layer is linked to predict the class and midpoint of each training picture chunk. Finally, the weighted fusion layer fuses the side outputs from various scales [21].

To train its Model, the candidate block screening network inputs the sampled image blocks. It then returns the positive sample's center offset and the category label for each sample block. In the test phase, the trained candidate block network is applied to all breast cancer tissue pathology images for an entire convolution operation to generate an anchor array, and each anchor is judged to screen out the candidate filament components of the whole picture. Since there are a lot of redundant estimates in this detection result, soft non-maximum suppression (Soft-NMS)

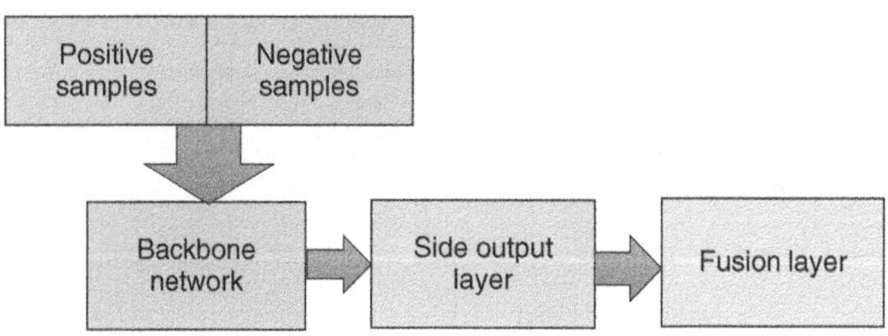

Figure 5.4 Candidate block network structure diagram.

[22] is used to remove redundancy. Unlike the previous non-maximum suppression, which roughly reduces the score of the detection box with a threshold more significant than the highest scoring detection box IOU value to zero, the introduction of Soft-NMS effectively alleviates this design defect of traditional NMS. The weight variable value is updated by linear and Gaussian weighting, which does not require retraining the original Model. The program is relatively easy to implement, and the amount of calculation will not increase. It can also be easily combined with the target detection algorithm, which has a specific improvement effect on detecting dense targets. Specifically, the coordinates of all detected cells are input, and the distance from each cell to all other cells is calculated. After traversing all cells, a distance threshold is given. The threshold here is set to 60 pixels, and save all cell indexes whose distance to the cell is less than the threshold. When the number of indexes is greater than 2, the coordinates of the point with the most significant score in the index are taken out, thereby removing redundancy. Finally, a candidate block of size 60×60 is extracted with the detected position as the center. The distance between each predicted cell position coordinate and the manually marked position coordinates in the database is calculated. If there is a distance within the range of 60 pixels, the detected candidate point is considered to be a mitotic cell and is given a label of "1"; otherwise, if the distance exceeds the range of 60 pixels, it is a non-mitotic cell and is given a label of "0"; then, the candidate blocks extracted from each pathological image and the labels determined are saved to facilitate the Training of the second network.

5.2.4 Verification Module—Task-Guided Radial Basis Network

The traditional radial basis function network (RBF Network) [23] is just a simple three-layer feedforward structure containing three layers: input, hidden (nonlinear), and output (linear). In most cases, the network input features are manually retrieved from the source image; as a result, fitting the data and extracting features are two separate operations. An improved feature representation for RBF networks can be obtained using CNN due to its remarkable capacity to learn hierarchical spatial features from raw data. Therefore, this chapter innovatively integrates feature extraction and RBF expression into a unified framework, which can not only learn high-level feature expression in a data-driven manner but also give full play to the advantages of the excellent approximation and generalization capabilities of the radial basis network. At the same time, the definition of the RBF center is combined with the task of mitotic cell identification. First, a typical convolutional network (without RBF) is used to classify and pre-train samples to obtain the initial weights of the network, which is used as a feature extractor to express the features of mitotic samples. Based on this, the standard K-means clustering algorithm is run to initialize the RBF center, and then the RBF is used to identify

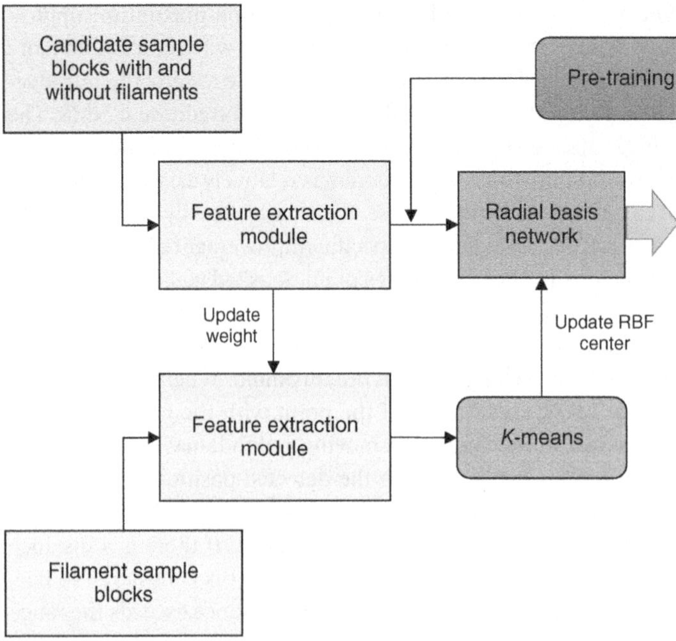

Figure 5.5 Verification module, task-guided radial basis network structure diagram.

the mitotic cells. The convolutional network is then used for classification, and the optimal radial basis center is determined by iterative clustering update. The work of [24] inspires this profound K-means clustering idea. The iteratively corrected RBF function center can more accurately express the morphological structure of diverse mitotic cells. The specific verification module network structure diagram is shown in Figure 5.5.

Specifically, for the feature extraction module, the filamentous or non-filamentous candidate block information retrieved from the initial candidate block network is fed into the feature extraction network for pre-training. After the feature extraction of the first four layers of convolutional modules in Resnet18, the maximum pooling layer is connected to obtain the features of size 1×1. Finally, two fully connected layers are constructed to reorganize and reduce the dimensions of the features, and classification deep learning is performed in the pre-training stage. Using all the filamentous samples in the training set, a series of filamentous sample features are obtained according to the pre-trained feature extractor; formula (5.1) shows that the filamentous sample blocks are clustered using the K-means technique. Each feature in the sample is assigned to the cluster whose center is nearest to it based on the computed distance from that feature to the cluster center. Following the assignment of all points, the cluster's center is recalculated by averaging all the cluster's points. Lastly, the clustering process is finished after the number of iterations has been reached and the cluster center has been determined. Alternatively, assigning points and updating the cluster center point are repeated until the change is minor.

$$\min_{C\in\mathbb{R}^{N}-n} \frac{1}{N}\sum_{n=1}^{N} \min_{y_n\in\{0.1\}} \|f_\theta(x_n) - Cy_n\|_2^2, \text{ s.t. } {y_n}^{\mathrm{T}}1_m = 1 \tag{5.1}$$

Among them, $f_\theta(x_n)$ represents the output feature of the CNN, each input x_n is associated with a label i_n between $\{0, 1\}^m$, is the feature's dimension, C is the radial basis center, and m is the number of cluster centers.

The second is the task-guided radial basis network. To better describe the diversity of mitotic cells in breast pathology images, this chapter adds a radial basis network layer after the Resnet18 network and uses different radial bases to express mitotic cells of a particular shape. There is a linear mapping from the space of the hidden layers to the space of the network's output, meaning that the center generated by RBF after K-means clustering determines this mapping relationship. Consequently, the network's output is the hidden units' outputs multiplied by a linear weight. The RBF network's node activation function uses the RBF, a Gaussian function, to estimate the distance between the input and the center, as shown in Formula (5.2). This contrasts with the multi-layer perceptron, which uses a linear function for its hidden node basis and a Sigmoid function for its activation function. The more similar, the closer. The larger the distance, the closer the output is to 1; conversely, the smaller the distance, the closer the output is to 0. The "local response" characteristic of the radial basis is used to describe the morphological characteristics of different mitotic samples effectively. The responses of different RBFs are linearly weighted to obtain the final prediction probability of mitotic cells, as shown in formula (5.3).

$$R(I_p - C_k) = \exp\left(-\frac{1}{2\sigma^2}\|I_p - C_k\|^2\right), \quad k = 1, 2, \dots, m \tag{5.2}$$

wherein I_p represents the input vector, that is, to which the CNN output is applied, C_k represents the cluster center, that is, the center of the kth Gaussian function, σ is the variance of the Gaussian function, which can be used to adjust the influence radius. The empirical formula is $\sigma = \frac{C_{\max}}{\sqrt{2m}}$, where $C_{\max}$ refers to the most excellent possible separation between cluster nodes, where m is the count of nodes in the cluster.

Finally, the center update module is used. The pre-trained weights are assigned to the clustering module after the candidate block input feature extraction module pre-trains the network. The clustering centers of different clusters are generated by clustering the mitotic cell sample blocks in the database. Then, the radial basis network centers are initialized using these clustering centers. Then, the candidate blocks are reclassified using the convolutional network with an RBF layer added, and the updated network weights are re-assigned to the clustering module for mitotic clustering to obtain new clustering centers. This new set of clustering centers is used to update the basis function center of the radial basis network in the next iteration to perform category judgment learning on mitotic cells. The network model gradually reduces the error through continuous iterative correction, further optimizes, and improves the detection performance. The network's final output is obtained by the weighted combination of different RBFs, as shown in formula (5.3), where Y represents the network's final output.

$$Y = \sum_{k=1}^{m} \omega_k R(I_p - C_k) \tag{5.3}$$

Similar to the first network testing stage, the final detection of this chapter is to detect mitotic cells by connecting the weights of the first network and the second network and applying the trained algorithm framework to the entire breast pathological tissue image.

5.2.5 Loss Function

In the candidate block network of the first stage, the sampled image blocks are trained for categories. This chapter adds a weight coefficient based on the cross-extraction loss function to deal with algorithm (5.4), which shows that there is an imbalance between the positive and negative samples due to weighting:

$$\begin{aligned} L_{\text{cls}}(\grave{l}; l) = &- \sum_{i} \left[\frac{N^-}{N^+ + N^-} l_i \log \grave{l}_i\right] \\ &- \sum_{i} \left[\frac{N^+}{N^+ + N^-} (1 - l_i) \log(1 - \grave{l}_i)\right] \end{aligned} \tag{5.4}$$

Among them, N^+ represents the number of positive sample image blocks; N^- represents the number of negative sample image blocks; l represents the ground truth of the input block using the binary label, that is, negative sample image blocks are marked with 0 and positive samples are marked with 1; l_i is the probability of prediction as a kernel in the range of [0, 1] obtained by Softmax classifier.

The position regression loss in the candidate block network is calculated only based on the positive samples and defined using the Euclidean L_2 norm, as shown in formula (5.5):

$$L_{\text{Reg}}(\bar{x}, \bar{y}; x, y) = -\sum_{i} 1(l_i = 1) \left[(x_i - \bar{x}_1)^2 + (y_i - \bar{y}_i)^2\right] \tag{5.5}$$

Among them, 1(.) is the regression indicator function, which depicts the candidate block with favorable results alone. (x, y) is the ground truth of the cell nucleus center offset recorded during the image block sampling process, and $(\bar{x}, \bar{y})$

is the estimated value of the center offset predicted by the candidate block selection network. Finally, the expression of the overall loss function calculated by combining the side output layer and the weighted fusion layer is shown in formula (5.6):

$$L(W) = \sum_{s} L_{\text{cls}}^{s}(\grave{l}; l) + L_{\text{cls}}^{\text{fuss}}(\grave{l}; l) + \gamma \left(\sum_{s} L_{\text{res}}^{s}(\bar{x}, \bar{y}; x, y) + L_{\text{reg}}^{\text{frase}}\left(\bar{x}, \bar{y}; x, y\right) \right) \tag{5.6}$$

Among them, W represents all parameters to be learned in the candidate block network; s represents the side output of different network layers; γ is a hyperparameter that balances this chapter's experimental design. This hyperparameter is set to 2 to represent the cost between regression and classification tasks. L_{cls}^{s} represents the category loss of the output layer on the kth side; L_{reg}^{s} stands for the function that determines the position loss on the sth side; $L_{\text{cls}}^{\text{fsse}}$ represents the category loss function of the weighted fusion layer; $L_{\text{reg}}^{\text{fuse}}$ represents the position loss function of the weighted fusion layer.

In the second stage of the task-guided radial basis network, to balance negative samples outnumber positive ones in the candidate blocks generated by the candidate block network, this chapter still designs to use formula (5.4) as the loss function in the pre-training part, while in the task-guided radial basis network, the Training uses the binary cross entropy loss function to perform category judgment deep learning on the candidate blocks, as shown in formula (5.7):

$$L(\bar{l}; l) = -\frac{1}{N} \sum_{i} [l_i \ln \bar{l}_i + (1 - l_i) \ln(1 - \bar{l}_i)] \tag{5.7}$$

where N is the number of samples; l is the binary labeled input block's ground truth, i.e. non-mitotic cells are marked with 0 and mitotic cells are marked with 1; $\bar{l}$ is the final predicted probability obtained by the weighted combination of different RBFs, i.e. the normalized result of the formula (5.3).

5.3 Evaluation and Results from Experiments

5.3.1 Experimental Setting

Here is the information regarding the experimental environment used for Training and testing in this chapter. Hardware configuration: CPU is Intel (R) Xeon (R) Gold5118; GPU is NVIDIATITANRTX; memory is 24 GB; software configuration: operating system is 64-bit Windows 10; Python3.7; PyTorch1.7. The candidate block network uses the Adam optimizer for algorithm optimization training, while the SGD optimizer is selected in the task-guided radial basis network. When training the two-stage network model, the epoch is set to 200, the initial deep learning rate is 0.001, and the Step LR deep learning rate adjustment strategy mechanism is used. An update is performed every eight epochs, the decay factor is 0.2, the momentum is set to 0.9, the second-stage network iteration update number is set to 100, and 50 small batch image blocks are randomly selected to make the model network training achieve the best performance.

5.3.2 Performance Evaluation Metrics

In the mitosis detection task, its performance is measured by the number of mitotic cells that can be accurately detected in breast cancer pathology images. According to ICPR 2014 and AMIDA 2013 competition standards, the accuracy of mitosis detection means that if the distance between the detected position and the marked center label is within 8μm (32 pixels), it is considered a true positive. The competition proposes that the accuracy of mitosis detection can be evaluated using recall, precision, and *F*-score. Among them, a detection model's performance measures are recall and precision. Recall is the ratio of actual mitotic cells detected by the Model. At the same time, precision is the ratio of all cells detected to the number of mitotic cells detected by the Model. The detection model is good if the accuracy and recall rates are well balanced. By summing the recall and precision rates harmonically, we get the *F*-score. The *F*-score is higher, assuming a near-par relationship between the recall and precision rates. The value is close to the average of recall and precision.

5.3.3 The Impact of Picture Block Size on the Efficiency of the Model

To verify the effect of the image block size used for classification and regression positioning on the model performance, this chapter sets the input positive and negative sample image block sizes in the candidate block network to 40×40, 50×50, 60×60, and 70×70, respectively. By counting the number of TP, FN, and FP detected by the candidate block network on the ICPR 2014 validation dataset, the detection performance of image blocks of different sizes is shown in Table 5.1.

Table 5.1 The number of TP, FN, and FP of image patches of different sizes on the validation set.

	40 × 40	50 × 50	60 × 60	70 × 70
TP	88	107	121	119
FN	64	45	31	33
FP	166	146	152	179

The experimental results show that the Model performs best when the image block size is 60 × 60 pixels. The image block of 40 × 40 pixels is too small to include all the features of mitotic cells, resulting in the loss of crucial position information for distinguishing mitotic and non-mitotic cells, causing a large number of missed detections in the candidate block network detection results, which is not conducive to the network training and detection performance of the subsequent second-stage verification module; the image block of 70 × 70 pixels is too large, so the positive sample image block contains too much useless information, which will lead to a large number of FNs, thus affecting the task-guided radial basis network for the final detection of mitotic cells; the number of TP, FN, and FP detected by the image block of 50 × 50 pixels and 60 × 60 pixels on the test set is similar. Still, through calculation, it is found that the Model's performance is higher when the image block of 60 × 60 pixels is input than that of the image block of 50 × 50 pixels. Therefore, this chapter finally chooses to collect the image block of 60 × 60 pixels as input for the Training of the candidate block network.

5.3.4 The Impact of Deep Supervision and Attention Mechanism on Model Performance

In this study, we performed ablation experiments of the CBAM attention module and deep supervision mechanism on the validation set divided by ICPR 2014 to verify the impact of both mechanisms on model performance. As you can see from Table 5.2, we compared the four schemes. By considering the prediction ability of different network scale levels, the deep supervision applied to the network model in this chapter improves the deep learning process by making the network's hidden layer more direct and transparent, compared to a traditional CNN without deep supervision. Each side output layer path simultaneously incorporates a CBAM attention module to generate an attention feature map that considers the channel and space dimensions. By multiplying the original input feature map with the information from the two feature maps, adaptive feature correction is used

Table 5.2 Ablation results of deep supervision mechanism and CBAM attention module on mitosis image patch classification.

Deep suspension mechanism	CBAM attention module	*F*-score
--	--	0.586
√	--	0.617
√	√	0.632

to recalibrate the channel features. Therefore, the network architecture of this chapter can find more candidate blocks of mitotic cells and obtain better detection performance than traditional deep neural networks.

5.3.5 The Impact of the Number of Cluster Centers on Model Performance

Finding the ideal number of cluster centers is the key to defining the RBF center, which is closely related to the classification performance of the RBF network. The objective function of the K-means clustering algorithm is to minimize the square error between the points of the particles and the samples. The degree of distortion is calculated as the total of the square distance errors between the places in each cluster where particles are located and the points where samples are taken. When there are fewer than the ideal number of cluster centers, increasing the value of m will cause each cluster to aggregate more tightly. Consequently, there is a marked reduction in distortion. When m approaches the ideal number of cluster centers, the return on increasing aggregation degree becomes smaller, and the distortion degree drops sharply. Then, as m rises, it climbs and eventually levels off. Based on this indicator, the number of cluster centers is set to 1, 2, 3, 4, 5, 6, 7, 8, and 9 in the K-means clustering model for repeated Training. And ten-fold cross-validation, the results are shown in Figure 5.6. Experiments show that when the cluster center is 3, the degree of distortion is greatly improved, and the clustering performance is better, so this article selects $m = 3$ as the number of cluster centers.

5.3.6 Experiments on ICPR 2014 Dataset

The samples in the database are proportionately split 4:1 between training and verification samples at random [25]. In the first stage of candidate block

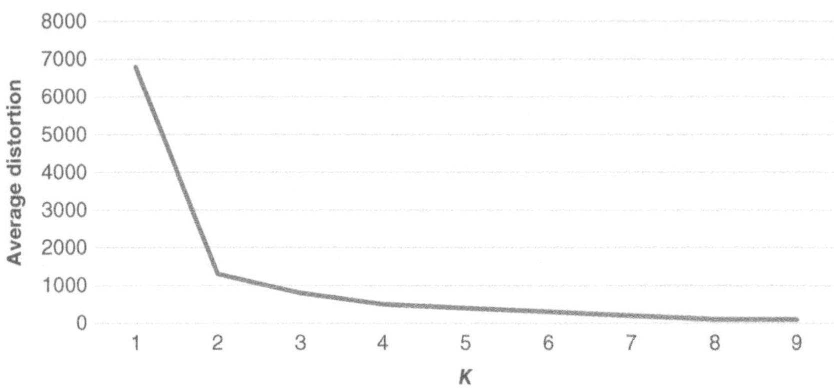

Figure 5.6 Determine the optimal K value using the elbow rule.

network training, positive and negative sample image blocks with a ratio of 1:2 are first collected. In this chapter, 9350 positive sample blocks containing the training data set contain cell types that undergo mitosis or non-mitosis, and 18,700 sample blocks that do not contain these types of cells. These image blocks have a dimension of 60×60. Whole image convolution is applied to the trained candidate block network, and about 17,136 mitotic cell candidate blocks with a size of 60×60 are statistically extracted. Then, they are input into the verification module for task-guided radial basis network training. The weights assigned by pre-training extract 60×60 candidate blocks from the training set. The K-means clustering algorithm generates the initialized RBF center for the 60-pixel sample blocks with approximately 495 mitoses. Finally, the iterative Training allows the network weights and centers to be continuously updated to find the optimal radial basis center, thereby achieving the final detection of mitotic cells.

To verify the effectiveness of the model performance, this chapter compares the current algorithms related to mitosis detection in breast pathology images on the ICPR 2014 dataset, as shown in Table 5.3.

Table 5.3 lists three quantitative performance evaluation indicators: precision, recall, and F-score. Experiments show that mitosis detection is an area where the suggested approach excels. Compared to other approaches, their performance is superior; specifically, their recall is 0.782, F-score is 0.685, and precision is 0.610. This proves the task-guided radial basis network is the best and most efficient method for detecting breast cancer mitotic cells. A green box represents TP, a yellow box represents FP, and a blue box represents FN; Figure 5.7 provides an example of the Model's detection findings on the ICPR 2014 validation set.

Table 5.3 Findings from various approaches on the ICPR 2014 validation set.

Algorithm	Precision	Recall	*F*-score
DeepMitosis [26]	—	—	0.572
MSSN [27]	0.379	0.617	0.470
SegMitos [28]	0.541	0.682	0.602
RCNN based [29]	0.530	0.660	0.585
Resnet-101 [30]	0.556	0.704	0.621
DIA+PMS [31]	0.581	0.691	0.631
OurNet	0.610	0.782	0.685

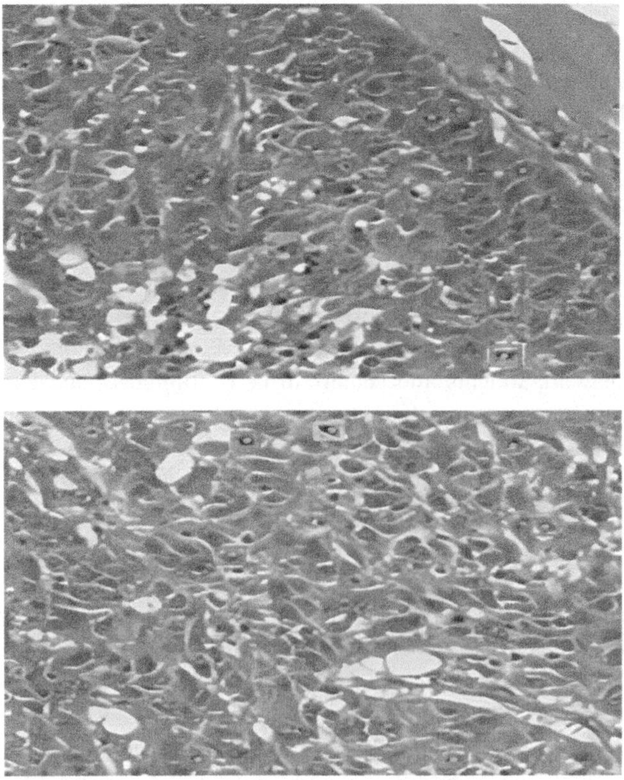

Figure 5.7 Detection results on two validation images at ICPR 2014.

5.3.7 Experiments on the AMIDA 2013 Dataset

The proposed Model is evaluated and verified on the AMIDA 2013 dataset. First, image blocks for classification and regression positioning are collected for the first stage of candidate block network deep learning. Then, candidate blocks are extracted and input into the second stage task-guided radial basis network to realize the detection of mitotic cells. The suggested method's performance on the AMIDA 2013 test set is compared to other approaches in Table 5.4. According to the experimental results of the AMIDA 2013 dataset, the suggested algorithm achieves the maximum F score of 0.728, recall rate of 0.707, and precision of 0.751. The Model's detection results on the AMIDA 2013 test set are illustrated in Figure 5.8, with TP, FP, and FN represented by green, yellow, and blue boxes, respectively.

Table 5.4 Findings from various approaches on the AMIDA 2013 set for validation.

Algorithm	Precision	Recall	*F*-score
LightweightDNN [32]	0.470	0.780	0.556
DeepResNet+HoughVoting [33]	0.547	0.686	0.609
SegMitos-random [34]	0.669	0.677	0.673
PartMitosis [35]	0.743	0.658	0.698
OurNet	0.751	0.707	0.728

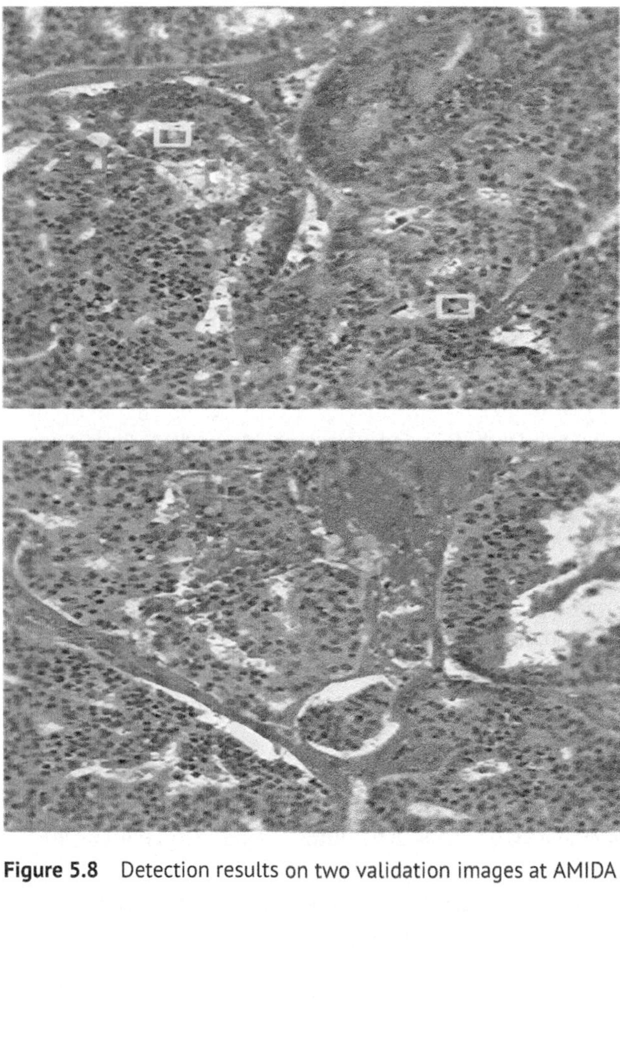

Figure 5.8 Detection results on two validation images at AMIDA 2013.

5.4 Conclusion

Aiming at the problem of morphological diversity of mitotic cells, this chapter proposes a novel automatic detection model for mitosis in breast pathological images. The overall scheme leverages multimodal imaging and molecular and optical approaches to enhance detection accuracy and consists of three parts: an image block acquisition model based on block learning, which is used to train the candidate block network; a detection model integrating a deep supervision mechanism, which utilizes multimodal imaging techniques to screen candidate blocks of mitotic cells; and a DNN model embedded with RBFs, enriched with molecular and optical approaches, to realize the final verification of mitotic cells. This method significantly improves the performance of mitotic cell detection using only weakly labeled datasets by combining the strengths of multimodal imaging for morphological analysis and molecular and optical approaches for structural refinement. Such integration provides compelling evidence for subsequent medical analysis and diagnosis. Experimental results show that the model constructed in this chapter can achieve a reasonable recall rate on both ICPR 2014 and AMIDA 2013 centroid-annotated datasets, resulting in fewer missed diagnoses. The use of multimodal imaging and molecular and optical approaches ensures robustness and adaptability, making this method significant for practical clinical applications.

References

1 Veta, M., Pluim, J.P.W., van Diest, P.J., and Viergever, M.A. (2014). Breast cancer histopathology image analysis: a review. *IEEE Transactions on Bio-Medical Engineering* 61 (5): 1400–1411. https://doi.org/10.1109/tbme.2014.2303852.

2 Ho, D.J., Herman Chui, M., Vanderbilt, C.M. et al. (2023). Deep interactive learning-based ovarian cancer segmentation of H&E-stained whole slide images to study morphological patterns of BRCA mutation. *Journal of Pathology Informatics* 14: 100160. https://doi.org/10.1016/j.jpi.2022.100160.

3 Wu, B. and Moeckel, G. (2023). Application of digital pathology and machine learning in the liver, kidney and lung diseases. *Journal of Pathology Informatics* 14: 100184. https://doi.org/10.1016/j.jpi.2022.100184.

4 Bowman, T.C., El-Shenawee, M., and Campbell, L.K. (2015). Terahertz imaging of excised breast tumor tissue on paraffin sections. *IEEE Transactions on Antennas and Propagation* 63 (5): 2088–2097. https://doi.org/10.1109/tap.2015.2406893.

5 Rajasekar, V., Vaishnnave, M.P., Premkumar, S. et al. (2023). Lung cancer disease prediction with CT scan and histopathological images feature analysis

using deep learning techniques. *Results in Engineering* 18: 101111. https://doi.org/10.1016/j.rineng.2023.101111.

6 Sauter, D., Lodde, G., Nensa, F. et al. (2023). Deep learning in computational dermatopathology of melanoma: a technical systematic literature review. *Computers in Biology and Medicine* 163: 107083. https://doi.org/10.1016/j.compbiomed.2023.107083.

7 Paul, A. and Mukherjee, D.P. (2015). Mitosis detection for invasive breast cancer grading in histopathological images. *IEEE Transactions on Image Processing: A Publication of the IEEE Signal Processing Society* 24 (11): 4041–4054. https://doi.org/10.1109/tip.2015.2460455.

8 Jiang, H., Diao, Z., Shi, T. et al. (2023). A review of deep learning-based multiple-lesion recognition from medical images: classification, detection and segmentation. *Computers in Biology and Medicine* 157: 106726. https://doi.org/10.1016/j.compbiomed.2023.106726.

9 Liu, L., Wang, Y., Zhang, P. et al. (2024). Collaborative transfer network for multi-classification of breast cancer histopathological images. *IEEE Journal of Biomedical and Health Informatics* 28 (1): 110–121. https://doi.org/10.1109/jbhi.2023.3283042.

10 Zhang, L., Jiang, S., Zhao, Y. et al. (2018). Direct regularization from co-registered contrast MRI improves image quality of MRI-guided near-infrared spectral tomography of breast lesions. *IEEE Transactions on Medical Imaging* 37 (5): 1247–1252. https://doi.org/10.1109/tmi.2018.2794548.

11 Sebai, M., Wang, T., and Al-Fadhli, S.A. (2020). PartMitosis: a partially supervised deep learning framework for mitosis detection in breast cancer histopathology images. *IEEE Access: Practical Innovations, Open Solutions* 8: 45133–45147. https://doi.org/10.1109/access.2020.2978754.

12 Thirusittampalam, K., Julius Hossain, M., Ghita, O., and Whelan, P.F. (2013). A novel framework for cellular tracking and mitosis detection in dense phase contrast microscopy images. *IEEE Journal of Biomedical and Health Informatics* 17 (3): 642–653. https://doi.org/10.1109/titb.2012.2228663.

13 Yang, H., Kim, J.-Y., Kim, H., and Adhikari, S.P. (2020). Guided soft attention network for classification of breast cancer histopathology images. *IEEE Transactions on Medical Imaging* 39 (5): 1306–1315. https://doi.org/10.1109/tmi.2019.2948026.

14 Aziz, S., Munir, K., Raza, A. et al. (2023). IVNet: transfer learning based diagnosis of breast cancer grading using histopathological images of infected cells. *IEEE Access: Practical Innovations, Open Solutions* 11: 127880–127894. https://doi.org/10.1109/access.2023.3332541.

15 Das, A., Devarampati, V.K., and Nair, M.S. (2022). NAS-SGAN: a semi-supervised generative adversarial network model for atypia scoring of

breast cancer histopathological images. *IEEE Journal of Biomedical and Health Informatics* 26 (5): 2276–2287. https://doi.org/10.1109/jbhi.2021.3131103.

16 Wang, H., Hao, X., Li, B. et al. (2024). A novel dataset and a two-stage mitosis nuclei detection method based on hybrid anchor branch. *Biomedical Signal Processing and Control* 87: 105374. https://doi.org/10.1016/j.bspc.2023.105374.

17 Su, Y.-T., Yao, L., Liu, J. et al. (2021). Spatio-temporal mitosis detection in time-lapse phase-contrast microscopy image sequences: a benchmark. *IEEE Transactions on Medical Imaging* 40 (5): 1319–1328. https://doi.org/10.1109/tmi.2021.3052854.

18 Xu, J., Xiang, L., Liu, Q. et al. (2016). Stacked sparse autoencoder (SSAE) for nuclei detection on breast cancer histopathology images. *IEEE Transactions on Medical Imaging* 35 (1): 119–130. https://doi.org/10.1109/tmi.2015.2458702.

19 Huh, S., Ker, D.F.E., Bise, R. et al. (2011). Automated mitosis detection of stem cell populations in phase-contrast microscopy images. *IEEE Transactions on Medical Imaging* 30 (3): 586–596. https://doi.org/10.1109/tmi.2010.2089384.

20 Li, X., Radulovic, M., Kanjer, K., and Plataniotis, K.N. (2019). Discriminative pattern mining for breast cancer histopathology image classification via fully convolutional autoencoder. *IEEE Access: Practical Innovations, Open Solutions* 7: 36433–36445. https://doi.org/10.1109/access.2019.2904245.

21 Li, Y., Xue, Y., Li, L. et al. (2022). Domain adaptive box-supervised instance segmentation network for mitosis detection. *IEEE Transactions on Medical Imaging* 41 (9): 2469–2485. https://doi.org/10.1109/tmi.2022.3165518.

22 Liu, M., Lanlan, H., Tang, Y. et al. (2022). A deep learning method for breast cancer classification in the pathology images. *IEEE Journal of Biomedical and Health Informatics* 26 (10): 5025–5032. https://doi.org/10.1109/jbhi.2022.3187765.

23 Han, J., Wang, X., and Liu, W. (2021). Contextual prior constrained deep networks for mitosis detection with point annotations. *IEEE Access: Practical Innovations, Open Solutions* 9: 71954–71967. https://doi.org/10.1109/access.2021.3079215.

24 Chen, J., Jiao, J., He, S. et al. (2021). Few-shot breast cancer metastases classification via unsupervised cell ranking. *IEEE/ACM Transactions on Computational Biology and Bioinformatics* 18 (5): 1914–1923. https://doi.org/10.1109/tcbb.2019.2960019.

25 Moussavi, F., Wang, Y., Lorenzen, P. et al. (2014). A unified graphical models framework for automated mitosis detection in human embryos. *IEEE Transactions on Medical Imaging* 33 (7): 1551–1562. https://doi.org/10.1109/tmi.2014.2317836.

26 Baran, P., Mayo, S., McCormack, M. et al. (2018). High-resolution X-ray phase-contrast 3-D imaging of breast tissue specimens as a possible adjunct to histopathology. *IEEE Transactions on Medical Imaging* 37 (12): 2642–2650. https://doi.org/10.1109/tmi.2018.2845905.

27 Song, H., Sasada, S., Kadoya, T. et al. (2024). Cross-correlation of confocal images for excised breast tissues of total mastectomy. *IEEE Transactions on Bio-Medical Engineering* 71 (5): 1705–1716. https://doi.org/10.1109/tbme.2023.3348480.

28 Beevi, K.S., Nair, M.S., and Bindu, G.R. (2017). A multi-classifier system for automatic mitosis detection in breast histopathology images using deep belief networks. *IEEE Journal of Translational Engineering in Health and Medicine* 5: 1–11. https://doi.org/10.1109/jtehm.2017.2694004.

29 Nie, W., Cheng, H., and Yuting, S. (2017). Modeling temporal information of mitotic for mitotic event detection. *IEEE Transactions on Big Data* 3 (4): 458–469. https://doi.org/10.1109/tbdata.2017.2723395.

30 Albarqouni, S., Baur, C., Achilles, F. et al. (2016). AggNet: deep learning from crowds for mitosis detection in breast cancer histology images. *IEEE Transactions on Medical Imaging* 35 (5): 1313–1321. https://doi.org/10.1109/tmi.2016.2528120.

31 Alom, M.Z., Aspiras, T., Taha, T.M. et al. (2020). MitosisNet: end-to-end mitotic cell detection by multi-task learning. *IEEE Access: Practical Innovations, Open Solutions* 8: 68695–68710. https://doi.org/10.1109/access.2020.2983995.

32 Pan, X., Lu, Y., Lan, R. et al. (2021). Mitosis detection techniques in H&E stained breast cancer pathological images: a comprehensive review. *Computers & Electrical Engineering: An International Journal* 91: 107038. https://doi.org/10.1016/j.compeleceng.2021.107038.

33 Jahanifar, M., Shephard, A., Zamanitajeddin, N. et al. (2024). Mitosis detection, fast and slow: robust and efficient detection of mitotic figures. *Medical Image Analysis* 94: 103132. https://doi.org/10.1016/j.media.2024.103132.

34 Roux, L., Racoceanu, D., Loménie, N. et al. (2013). Mitosis detection in breast cancer histological images an ICPR 2012 contest. *Journal of Pathology Informatics* 4 (1): 8. https://doi.org/10.4103/2153-3539.112693.

35 Lashen, A., Toss, M.S., Alsaleem, M. et al. (2022). The characteristics and clinical significance of atypical mitosis in breast cancer. *Modern Pathology* 35 (10): 1341–1348. https://doi.org/10.1038/s41379-022-01080-0.

6

Advancements in Biomedical Imaging Using Fluorescence and Bioluminescence

Ashish Kashyap[1], Manju Jakhar[2], Nidhi Rani[3], and Thakur Gurjeet Singh[3]

[1] *Department of Pharmacology, Bhojia Dental College and Hospital, Nalagarh, Baddi, Himachal Pradesh, India*
[2] *School of Pharmacy and Emerging Sciences, Baddi University of Emerging Sciences & Technology, Baddi, Himachal Pradesh, India*
[3] *Chitkara College of Pharmacy, Chitkara University, Punjab, India*

6.1 Introduction

Biomedical imaging has exponentially transformed our perspective of biological systems. Fluorescence and bioluminescence imaging (BLI) in particular are crucial imaging technologies that enable visualization at the molecular level, providing critical information related to cell function, disease progression, and therapeutic interventions [1].

Fluorescence images are produced by exciting fluorophores with light from external light sources, the fluorophores emit light when excited, giving details to the cellular components in real time. On the other hand, BLI utilizes light in the visible spectrum produced through biochemical reactions requiring luciferase enzymes and substrates. Fluorescence imaging is less sensitive than BLI but has higher spatial resolution and, when combined with fluorescent proteins, can enable accurate imaging of cellular components; whereas body bioluminescence imaging is highly sensitive, capable of detecting signals from very low-abundant species with minimal background interference [2–4].

Advances in these modalities have enabled them to interrogate biological systems deeper and with a higher accuracy and precision. This chapter investigates the state of the art, virtues, and hurdles of fluorescence and BLI, emphasizing their revolutionary possibilities in biomedical science [5].

Biomedical Imaging Technology: Signal Processing Strategies and Innovations, First Edition.
Edited by Ayush Dogra, Shalli Rani, and Ankita Sharma.

6.2 Advancements in Imaging Bioluminescence

6.2.1 Advances in Bioluminescence Imaging

BLI is a unique non-invasive, dynamic imaging technique to visualize biological activities in real time without the use of homologous light sources. An important step forward is BREAKFAST, a bioluminescent probe that allows for dynamic four-color imaging via resonant energy transfer. This innovation not only broadens cellular-scale imaging capabilities, but also paves the way for subsequent breakthroughs in biomedical applications [6–8].

Optimizations in natural bioluminescent systems have enhanced interactions between luciferase and luciferin such that biological processes can be timed in real time, making the technique of BLI become an unavoidable tool in onco-areas and neuroscience study [9]. In comparison, BLI can image the brain deeper since it does not have limitations owing to photon scattering, unlike in fluorescence [10].

Despite its benefits, there are disadvantages in BLI, such as a lower quantum yield that demands a long exposure time. New advances in the creation of new luciferase–luciferin systems expand the existing utility of this modality in imaging and add assurance to its promise for serving as a real-time biological monitor [11].

6.2.2 Fluorescence Imaging Challenges

Fluorescence is powerful but limited by phototoxicity and autofluorescence. High-intensity light can give rise to a phenomenon known as phototoxicity, destroying biological samples and changing activities within cells. Autofluorescence, the intrinsic light emission of biological tissues, frequently hampers signals from fluorescent probes preventing image clarity [12].

New tomographic methods are rapidly developing to achieve reliable quantitative analysis, including in the context of cancer. Such advancements will allow for more reliable and precise fluorescence imaging and address the limitations of existing techniques for deep-tissue imaging of biological phenomena [13].

6.2.3 Recent Innovations in Imaging Technologies

Recent breakthroughs in fluorescence and BLI, especially NIR-II technologies, have revolutionized biomedical imaging. NIR-II fluorescence imaging at 1000–1700 nm shows increased tissue penetration and fidelity, allowing for deeper visualization of biological processes. This development overcomes barriers such as absorption and scattering in tissues, enabling promising use cases in biomedical settings [14].

This was further aided by genetically encoded and synthetic organic fluorophores, as well as fluorescent nanoparticles, which enabled RNA interactions and kinase activity to be studied with unprecedented precision [15]. Similarly, BLI technologies employing NIR-II emissions would be used for further signal-to-noise ratios (SNRs) and spatial resolution, which is critical for the imaging of tumor metastasis. Improved luciferase–luciferin systems and their conjugation to the optogenetic platforms have redefined BLI by increasing the sensitivity of this tool and making it more suitable for real-time applications. These improvements have broadened the ability of biological pathways in visualization and manipulation [16, 17].

6.3 Key Innovations in Bioluminescence Imaging (BLI)

BLI is a new imaging modality in biomedical imaging, which provides higher sensitivity and the capability to visualize biological events in real time, in vivo. Since it is non-invasive and can be used to study cellular and molecular dynamics, this technique has become an essential tool for preclinical research. Although BLI has several merits, the low quantum yield and qualitative restrictions of BLI are serious limitations in the use of this technique for quantitative assays [18, 19].

6.3.1 Recent Advances

6.3.1.1 Luciferase-Loaded Nanoparticles

Calcium phosphate nanoparticles containing luciferase afford more lasting bioluminescence that can typically allow prolonged imaging of tumor model systems. This boosts temporal resolution vital to long-term studies [20].

6.3.1.2 Synthetic Bioluminescent Reactions

The development of synthetic luciferin–luciferase pairs has improved photon detection. In addition, these pairs facilitate high-speed imaging of refractile motion of animals in their natural environment, which gives improved understandings of dynamic physiological processes [21].

6.3.1.3 Bioluminescent Reporters

Optimization of different components of the reporter system such as the choice of promoter, codon optimization, of the molecular architecture of the fusion proteins has provided with high SNR. The most significant of such advancements is utilizing these advances for tracking the fate of stem cells in regenerative medicine and tissue engineering [22].

6.3.1.4 Bacterial Bioluminescence

The application of engineered bioluminescent bacteria has been characterized as new approaches toward real-time imaging of infections, tumor-directed therapies, probiotic associations, and drug testing. The integrated and heterogeneous biosensing platform has also been utilized for whole-cell biosensing, broadening the potential biological sensing applications [23].

6.3.1.5 Applications and Future Directions

- BLI has also allowed for the non-invasive monitoring of stem cell behavior, which is crucial for the enabling of precision medicine by being integrated into both therapy and tissue engineering strategies [24].
- Novel advances in the bacterial BLI technique are likely to streamline their usage in various fields, for example in oncology, microbiology, and pharmacology. As much as these advancements are strides by themselves, addressing challenges such as low quantum yield and heterogeneity of biological systems are essential to harnessing their full clinical potential for BLI [25, 26].

6.4 Limitations of Bioluminescence Imaging (BLI)

BLI has become a fundamental tool in biomedical research as it allows monitoring in real time without invasive usurpations. Yet, there are limitations emerging from the technology that may render it less effective, less accurate, and less acceptable for research or clinical practice, despite the great implications to the field. Recognizing the challenges is crucial for researchers in setting experiments and interpreting data [25, 27].

6.4.1 Depth Limitations

The penetration of tissue by various methods, including confocal microscopy, is restricted to a depth of 1–5 MFPs in scattering media [28]. BLI, however, has reached cellular resolution at 5–10 MFPs in scattering media. It does not, however, offer the depth of imaging like two-photon microscopy. But the brightness of bioluminescent indicators greatly impacts the depth of imaging. As tissue depth increases, the signal intensity rapidly decreases, which limits the ability to visualize deeper tissues, such as internal organs or tumors, where high-resolution imaging is often needed [29, 30].

6.4.2 Variation in Outputs

Labeling these sources of variability from bioluminescent reporters and their biochemical dependencies result in different experimental outputs:

- **ATP Dependent:** The performances of the reporters are quite different. This is particularly the case with the dynamic conditions that characterize most settings where there are differences in ATP concentrations. AP-dependent luciferases usually produce varied outputs under these dynamic conditions [31].

- **Stability in the Serum and Molecular Size:** The molecular size and stability of reporters in the serum adds another level to variability, thus further limiting the reproducibility of results of an experiment [32].

6.4.3 Limitations to Quantitative Precision

This very qualitative nature makes it impossible for BLI to be accurate quantitatively in outputting data. The process has a hard time measuring precisely through:

Systematic Errors: Changes in luminescence intensity measurements caused by the position of the subject and depth of the bioluminescent source. Identifying multiple luminescent sources at greater depths is a challenge. Although bioluminescence tomography (BLT) provides promising avenues, these techniques are limited by complexity [33, 34].

6.4.4 Other Major Limitations

- **Low Quantum Yield:** Many bioluminescent reactions have low quantum yield requiring long exposure times to obtain sufficient signal. This results in loss of resolution and might mean bigger collection wells to collect enough light [35].
- **Signal-to-Noise Ratio (SNR):** BLI generally has an SNR below 1, which severely limits the detection of weak signals. Signals clarity and accuracy are constantly improving with advanced enhancement techniques, such as deep learning algorithms [36].
- **Qualitative:** Being inherently qualitative, BLI presents challenges in interpreting specific quantitative data, making it unsuitable for thorough quantitative investigations of biological processes [37] (Table 6.1).

BLI has been shown to be a very sensitive, non-invasive imaging modality, but the caveats discussed must be considered when designing experiments that utilize more than one imaging modality within the same experiment. The aforementioned challenges can be addressed by a multi-approach strategy including bioluminescent reporter optimization, enhancements in imaging conditions, and the use of advanced computational tools that may improve the utility of BLI in preclinical/translational research settings [43, 44].

6.5 Evolution of BLI Technology

Advances in BLI technology have overcome many of these intrinsic limitations and been translated for more quantitative, sensitive, and quantitative applications

Table 6.1 A summary table of the limitations of BLI.

Sr. No.	Limitation	Explanation	Ref.
1	Qualitative nature	Mostly qualitative, so hard to take precise measurements.	[38]
2	Low quantum yield	Needs longer exposure times so image loses quality and sharpness.	[39]
3	Depth constraints	The depth penetration is limited in comparison to the advanced microscopy techniques.	[40]
4	Signal-to-noise ratio	Low SNR requires sophisticated enhancement techniques to suppress interference.	[39, 41]
5	Inconsistency in results	Bioluminescent reporters and their dependencies causing variability in outcomes.	[42]

to biomedical research over the past 20 years. Recent advancements in enzyme engineering, substrate design, and photon-detection systems have significantly elucidated the application of BLI as a promising and versatile tool for non-invasive kinetic monitoring in real-time biological processes [45, 46].

6.5.1 Enhanced Luminescent Units

Bioluminescence systems have been significantly enhanced by genetic engineering and chemical modifications:

- **Genetically Modified Enzymes:** The production of improved luciferases with higher light intensities and red-shifted light emitters substantially improved the ratio of signal over background noise in in vivo imaging. The optimized enzymes facilitate more reliable analytical data in deep-tissue imaging by significantly reducing light scattering [47].
- **Advanced Substrates:** Novel luciferin substrates have been designed, leading to bioluminescent traces with improved biodistribution and sensitivity. BRET and other methods for amplifying luminescence intensity may provide better imaging of molecular interactions [48].

6.5.2 Advanced Imaging Methods

Novel imaging methods have expanded the utility of BLI for diagnostics and research substantially:

- **A blue LASER Imaging (BLI) in Endoscopy:** This state-of-the-art technique significantly improves the visualization of the mucosal structures, with small lesions, at the gastrointestinal field [49].

Table 6.2 Variability improvements in bioluminescence imaging.

S. No.	Variable	Advancements	Ref.
1	Enzymes	Stability and sensitivity-enhancing enzyme development	[52]
2	Substrates	Novel substrates with improved biodistribution and sensitivity	[53]
3	Photon detectors	Photon detectors for 3D characteristics with excellent sensitivities	[54]

- **Non-Invasive Cancer Imaging:** The newly developed BLI technologies are applied to cancer research to study cancer progression dynamics and drug responses. Those studies uncover mechanism of malignancy progression and treatment efficacy, and clarify important biological insight [50].

6.5.3 Improvements in Photon Detection

6.5.3.1 High-Sensitivity Photon Detectors

New photon detectors will allow for three-dimensional acquisition, achieving better spatial separation of bioluminescence signals. These technologies allow for a more accurate tracking of cellular and molecular kinetics [51] (Table 6.2).

6.6 Applications of Bioluminescence Imaging

BLI has emerged as a strong strategy in biomedical research since it provides the possibility of non-invasive real-time monitoring with unprecedented sensitivity and resolution. With higher SNR, free of phototoxicity and photobleaching resistance, it is more advantageous than techniques based on fluorescence. In recent years, a number of breakthroughs have helped its application expand into multiple areas of research and medicine [55, 56].

6.6.1 Gene Expressions and Protein Localizations

BLI has become an essential tool for the study of gene expression and protein localization:

6.6.1.1 Multicolor Auto-Bioluminescence Systems

Systems such as Nano-lanternX (NLX) enable the measurement of gene expression and protein dynamics in a single living cell by multiplexed imaging.

NLX is compatible with bacterial, mammalian, and plant systems, allowing for detailed single-cell imaging and the assessment of complex biomolecular interactions [57, 58].

6.6.2 Tumor Imaging

In cancer research, BLI has become indispensable for tumor biology investigation and therapeutic intervention evaluation.

6.6.2.1 Long-Term Imaging with Nanoparticles

The luciferase-expressing luciferase-loaded calcium phosphate nanoparticles retain their luciferase activity in vivo for several weeks. This provides real-time orthotopic breast tumor imaging to offer in-depth understanding of tumor growth and metastasis [59, 60].

6.6.3 Optogenetic Biosensing

Combining BLI with optogenetics has been a significant advancement for biosensing and controlling cells.

6.6.3.1 Bioluminescence-Induced Optogenetic Biosensors

Genetic engineering of pairs of luciferases–luciferin enables the study of physiological events and the regulation of signaling pathways. These biosensors allow real-time monitoring of cellular processes, and in doing so, reveal potential therapeutics [61, 62] (Figure 6.1).

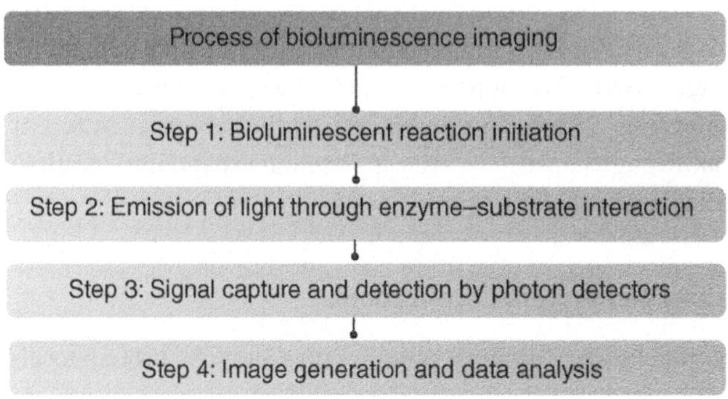

Figure 6.1 Process of bioluminescence imaging Adapted from [64].

6.6.4 Biomedical Research and Diagnostics

This enables BLI to have a wide range of applications in diagnostics and therapeutic research.

6.6.4.1 Studies of Infectious Disease and Compounds for Treatment

BLI enables noninvasive monitoring of the activity of pathogens, effectiveness of drugs, and response to therapies. It improves accuracy in biomedical research by providing real-time monitoring capabilities [64].

6.6.4.2 Challenges and Direction for the Future

BLI has several advantages associated with it, but it comes with some challenges also:

- The exogenous nature of luciferase expression and a high turnover rate of luciferin can prove to be disadvantages for products of longer studies [65].
- Photon absorption and scattering effect may reduce the signal intensity in deep-tissue imaging [66].

With luciferase engineering, substrate development and imaging modalities advancing to expand the utility of BLI in scientific and clinical settings, limitations in the modality have been overcome [67].

6.7 Innovations in Fluorescence Imaging

Fluorescence imaging is the bedrock of contemporary biomedical imaging, providing high-resolution snapshots of cellular structures and dynamic processes. Improvements emphasizing improved accessibility, accuracy, and functionality by means of miniaturization of fluorescent probes, computational photography and state-of-the-art imaging technologies have addressed these challenges recently. These advances are transforming the field, enabling researchers to explore complex biological pathways in unprecedented detail [68, 69].

6.7.1 Miniaturized Fluorescent Probes

Advancements in miniaturized fluorescent probes have revolutionized cellular and molecular imaging:

- **Small Probes:** New small fluorophores, including benzene-based fluorophores and fluorescent nucleobases. With both good sensibility and low interference in biological functions, this class of molecules has been termed optimal for in vivo applications [70, 71].

- **General Applications:** These probes enable true labeling of nucleic acids and proteins without interfering with their biological activities, thus permitting non-invasive imaging of cellular processes [72].

6.7.2 Computational Photography in Surgery

The hybridization of computational photography into the fluorescence imaging scales and redefines surgical and diagnostic application cases:

- **Economic Systems:** Computational photography new approaches have enabled the development of low-cost fluorescence imaging systems, hence making it accessible, especially to resource-poor regions [73].
- **Improved Visualization:** This technology allows for better visualization of tissues and blood vessels in real time during surgery, thereby improving precision and reducing complications [74].

6.7.3 Advanced Imaging Methods

New methods for fluorescence imaging are expanding the limits of resolution and functionality:

- **Light Sheet Fluorescence Microscopy (LSFM**): LSFM allows for high-resolution 3D imaging of biological specimens. It has been an indispensable tool for biomedical research due to its capacity to obtain high-fidelity information of disease mechanisms and therapeutic effects [75].
- **Fluorescence Molecular Tomography (FMT):** FMT has increasingly been applied in tumor diagnosis and therapeutic evaluations [76]. However, the latest trends, particularly with regard to deep learning-based image reconstruction, have significantly increased its accuracy and the possibility of diagnosis [77].

6.7.3.1 Challenges and Future Directions

Although these inventions are excellent progressions from former techniques, there is still the challenge of standardizing each of them and also implementing their application in clinical practice, especially in limited resource environments [78]. There is an ongoing process of streamlining these technologies to be embedded into everyday biomedical workflows to make fluorescence imaging more widely accessible. Contextualizing these breakthroughs within the broader field of fluorescence imaging emphasizes its integral contribution to contemporary biomedical research and clinical diagnostics [79]. Fluorescence

Table 6.3 Fluorescence imaging limitations.

S No.	Limitation	Explanation	Ref.
1	Phototoxicity and autofluorescence	High intensities cause sample damage and noise	[81, 84]
2	Complexity in quantification	Overlapping spectra complicate measurement	[85]

enables researchers to probe cellular mechanisms at unparalleled resolution; as a result, innovation is continuously expanding our understanding of disease and developing diseases from prevention to treatment [80].

6.7.3.2 Fluorescence Imaging: Limitations

- **Phototoxicity and Autofluorescence:** The strong excitation light intensity can cause phototoxicity that affects biological samples by causing cellular damage, thereby affecting the viability of the sample. Autofluorescence—unwanted fluorescence from the sample itself—can also cause complications in signal interpretation and data analysis [81–83].
- **Challenge Facing Quantitative Analysis:** Overlapped emission spectrum and variable concentrations of probe solutions, which usually makes quantitation difficult. Thus, such situations lead to misquantification and draw very misleading conclusions when using the collected experimental data for a conclusion (Table 6.3).

6.8 Advances in Fluorescence Imaging Technology

6.8.1 From Computational Photography to Fluorescence Imaging

A novel application of computational photography principles to achieve a successful reduction in the complexity and size of the optical design of the fluorescence imaging systems, making these affordable and accessible systems. This method leveraged off-the-shelf components combined with advanced algorithms to attain real-time, high-quality imaging comparable to that provided by commercial system. 5 mm penetration depth and sub- millimeter resolution system: 5 mm penetration depth and sub-millimeter resolution system; it may aid visualization of structures within the surgical field in low-resource countries [86, 87].

6.8.2 Near-Infrared Fluorescence Imaging in Cancer Diagnosis

The novel application of the near-infrared (NIR) fluorescence imaging technique has achieved the aim for non-specific fluorescent agents and targeted tracers toward supporting fluorescence-guided surgery (FGS) [88, 89].

This innovation has aided in identifying lymphatic vessels and nodes, thus helping to establish perfusion as well as recognize metastases in patients undergoing surgical interventions [90]. This technology assists surgeons in distinguishing lymphatic vessels and nodes, assessing perfusion, and visualizing metastases intraoperatively, although it remains an auxiliary method rather than a primary diagnostic tool [91].

6.8.3 Advances in Fluorescence Molecular Tomography (FMT)

- FMT is an invasive non-optical imaging technique with a high sensitivity and selectivity that can be helpful in performing tumor diagnosis and therapeutic evaluation [92].
- Focusing on the quality of reconstruction images for FMT, methods of deep learning and optimized algorithms were proposed in order to challenge the problems connected with photon scattering and the ill-posed questions of inverse problems [93].

6.8.4 Small-Molecule Probes in Bioimaging

Fluorescent small-molecule probes are developed with greater sensitivity and resolution than other imaging methods, the orderly assembly of imaging probe improved leading to better imaging performance as a result of better anti-diffusion and photostability capabilities [94]. These probes are necessary tools for acquiring cellular biochemical information, and have many applications in bioimaging and biomedical research [95].

6.8.5 Light Sheet Fluorescence Microscopy (LSFM)

LSFM has transformed three-dimensional fluorescence imaging through high spatiotemporal resolution and increased acquisition speed [96]. Recent advancements in LSFM architecture and computational analysis have broadened its utility in characterizing biological structures and disease manifestations, thereby assisting novel diagnostic and treatment paradigms [96].

Such advancements have even increased the power of fluorescence imaging but have raised other important challenges, such as the requirement of highly selective fluorescent molecular labels for detection of specific targets and uptake of these technologies into normal clinical practice [97]. Fluorescence imaging technologies have shown enormous promise in many applications, but ongoing research and development will be needed to meet these challenges and unlock their full potential [98].

6.9 Comparative Analysis of Bioluminescence and Fluorescence Imaging

Bioluminescence and fluorescence imaging: comparative aspects. Although both approaches are powerful tools for imaging cellular processes, they are different in terms of sensitivity, use case, and imaging methods.

6.9.1 Sensitivity and the Strength of the Signal

- **Bioluminescence Imaging (BLI):** This uses luciferase enzymes that produce light, often providing higher sensitivity for detecting fewer cells in vivo, especially in tumor models [99].
- **Fluorescence Imaging:** Tends to provide reasonable signal in greater depths of tissue as observed with NIR fluorophores that can more effectively label metastatic tumors within bone [100, 101].

6.9.2 Application and Versatility

- **Bioluminescence:** Typically used for reporting gene expression and cellular biology, but has also been found useful in studies of cancer biology and metabolism [102].
- **Fluorescence:** This offers different imaging modalities (e.g., fluorescence tomography, reflectance imaging), thus enabling 3D localization and quantification [103].

6.9.3 Hybrid Methods

The emergence of hybrid modalities has significantly enhanced the ability to explore complex biological processes. For instance, the BREAKFAST platform offers the possibility of simultaneous bioluminescence and fluorescence imaging, thus making it possible to monitor in real-time cellular dynamics [104].

For low-abundance targets, bioluminescence can be very sensitive; however, for deeper tissue evaluation, fluorescence imaging is a sturdy signal. The sum of both methods might potentially lead toward a holistic approach concerning biological processes [105] (Table 6.4).

Table 6.4 The comparison of the bioluminescence and fluorescence imaging techniques.

S. No.	Aspect	Bioluminescence imaging	Fluorescence imaging	Ref.
1	Sensitivity	Low background noise and high sensitivity	High sensitivity, but common background interference	[106]
2	Resolution	Moderate	Very high, in many techniques (e.g., STED, etc.)	[107]
3	Quantification	Difficult because qualitative nature	Complex (due to overlapping emission spectra)	[108]
4	Application	Useful in cancer research, infectious disease	It's used broadly in both cell biology and molecular diagnostics	[109]

6.10 Emerging Trends in Imaging Technological Development

Imaging technologies are developing toward wider applications with the development of innovative approaches to exploit fluorescence and BLI, as traditional methods have limitations [110]. For example, NIR-II imaging is a hot area to develop a new fluorescence technique that can penetrate deeper into the tissue with higher fidelity. The NIR-II imaging enables biomedical applications including bioimaging, biosensing, and therapeutic interventions [111].

6.10.1 Challenges and Suggestions

Advanced imaging techniques face both technical and biological challenges and if successful counts on these success. Areas to address include:

- **Novel NIR-II Fluorophores:** Imino quinoids have been identified as a class of fluorophores with longer wavelengths, high extinction coefficient, and high fluorescence quantum yield [112].
- **Significant Development of Quantification Methods:** This will be needed to overcome the limitations of signal derived from both fluorescence and BLI [113].
- **Fluorophore Design and Excitation Source:** New approaches that reduce phototoxicity and autofluorescence can make the imaging process safer and improve the accuracy of the images obtained [114].

6.11 Conclusion

Summary: The development of fluorescence and BLI technologies has greatly enhanced our ability to study biological events on a molecular scale. These have been significantly boosted by two key innovations: tailored bioluminescent systems and high-performance fluorescent ligands.

Super-resolution microscopy and fluorescence lifetime imaging (FLIM) are among the emerging techniques to overcome the phototoxicity and autofluorescence that limit the application of fluorescence imaging, which has a spatial resolution better than, for example, nuclear imaging. In contrast, BLI, which is characterized by low background interference and high sensitivity, has advanced in the real-time monitoring of biological processes, especially for oncology and infectious disease research. Still, despite the various advancements above, drawbacks such as low quantum yield, and qualitative limitations still pose hurdles for quantitative applications.

The chapter also discusses the future scope for these imaging technologies, especially through innovations such as NIR-II imaging and advanced quantification techniques that will help address the current limitations and expand their use in research and clinical settings. Such multimodal imaging approaches could provide further improved understanding of complex biological processes, eventually affording novel strategies in diagnostics and treatment in clinics. These results show that the imaging techniques discussed here have achieved only partial use, and more research and development will be necessary to fully utilize their capabilities.

References

1 Dickson, P.V., Hamner, B., Ng, C.Y.C. et al. (2007). in vivo bioluminescence imaging for early detection and monitoring of disease progression in a murine model of neuroblastoma. *Journal of Pediatric Surgery* 42 (7): 1172–1179.

2 Chen, X. (2013). Visualizing the location and the dynamics of gene expression in living animals through bioluminescence imaging. In: *Methods in Cell Biology*, vol. 113 (ed. P.M. Conn), 39–49. Academic Press.

3 Petryayeva, E., Russ Algar, W., and Medintz, I.L. (2013). Quantum dots in bioanalysis: a review of applications across various platforms for fluorescence spectroscopy and imaging. *Applied Spectroscopy* 67 (3): 215–252.

4 Mashanov, G.I., Tacon, D., Knight, A.E. et al. (2003). Visualizing single molecules inside living cells using total internal reflection fluorescence microscopy. *Methods* 29 (2): 142–152.

5 Liu, G. and Liu, Z. (2024). Fluorescence imaging. In: *Computational Optical Imaging: Principle and Technology* (ed. Z. Liu, X. Zhou, and S. Liu), 297–335. Singapore: Springer Nature Singapore.

6 Placone, J.K. and Engler, A.J. (2018). Recent advances in extrusion-based 3D printing for biomedical applications. *Advanced Healthcare Materials* 7 (8): e1701161.

7 Cosco, E. (2020). *Polymethine Fluorophores for in vivo Shortwave Infrared Imaging*. University of California, Los Angeles.

8 Contag, C.H. and Bachmann, M.H. (2002). Advances in in vivo bioluminescence imaging of gene expression. *Annual Review of Biomedical Engineering* 4 (1): 235–260.

9 Huang, Y., Liu, T., Huang, Q., and Wang, Y. (2024). From organ-on-a-chip to human-on-a-chip: a review of research progress and latest applications. *ACS Sensors* 9 (7): 3466–3488.

10 Kim, M., Panagiotakopoulou, M., Chen, C. et al. (2023). Micro-engineering and nano-engineering approaches to investigate tumour ecosystems. *Nature Reviews Cancer* 23 (9): 581–599.

11 Sun, N., Lee, A., and Wu, J.C. (2009). Long term non-invasive imaging of embryonic stem cells using reporter genes. *Nature Protocols* 4 (8): 1192–1201.

12 Prathap, N., Balla, P., Shivakumar, M.S. et al. (2023). Prosopis juliflora hydrothermal synthesis of high fluorescent carbon dots and its antibacterial and bioimaging applications. *Scientific Reports* 13 (1): 9676.

13 Yoon, S., Cheon, S.Y., Park, S. et al. (2022). Recent advances in optical imaging through deep tissue: imaging probes and techniques. *Biomaterials Research* 26 (1): 57.

14 Li, C. and Wang, Q. (2019). Advanced NIR-II fluorescence imaging technology for in vivo precision tumor theranostics. *Advanced Therapeutics* 2 (9): 1900053.

15 Laal, M. (2013). Innovation process in medical imaging. *Procedia-Social and Behavioral Sciences* 81: 60–64.

16 Frangioni, J.V. (2008). New technologies for human cancer imaging. *Journal of Clinical Oncology* 26 (24): 4012–4021.

17 Fan, W., Tang, W., Lau, J. et al. (2019). Breaking the depth dependence by nanotechnology-enhanced X-ray-excited deep cancer theranostics. *Advanced Materials* 31 (12): e1806381.

18 Feng, J., Qin, C., Jia, K. et al. (2011). Bioluminescence tomography imaging in vivo: recent advances. *IEEE Journal of Selected Topics in Quantum Electronics* 18 (4): 1394–1402.

19 Qin, Y., Shen, J., Qin, Y. et al. (2024). Research progress on the application of aptamer optimization technology and portable sensing devices in food safety detection. *Critical Reviews in Food Science and Nutrition* 1–33.

20 Saito, K., Chang, Y.F., Horikawa, K. et al. (2012). Luminescent proteins for high-speed single-cell and whole-body imaging. *Nature Communications* 3 (1): 1262.

21 Alonso-Roman, R., Mosig, A.S., Figge, M.T. et al. (2024). Organ-on-chip models for infectious disease research. *Nature Microbiology* 9 (4): 891–904.

22 Ganesh, R.B. and Maerkl, S.J. (2022). Biochemistry of aminoacyl tRNA Synthetase and tRNAs and their engineering for cell-free and synthetic cell applications. *Frontiers in Bioengineering and Biotechnology* 10: 918659.

23 Xu, X., Chueng, S., Jia, X. et al. (2024). Trends in organ-on-a-chip for pharmacological analysis. *TrAC Trends in Analytical Chemistry* 180: 117905.

24 Luo, T., Zhou, T., Zhao, Y. et al. (2018). Multiplexed fluorescence lifetime imaging by concentration-dependent quenching. *Journal of Materials Chemistry B* 6 (13): 1912–1919.

25 Zambito, G., Chawda, C., and Mezzanotte, L. (2021). Emerging tools for bioluminescence imaging. *Current Opinion in Chemical Biology* 63: 86–94.

26 Jiang, T., Bai, X., and Li, M. (2024). Advances in the development of bacterial bioluminescence imaging. *Annual Review of Analytical Chemistry* 17: 265–288.

27 Özkan, A., LoGrande, N.T., Feitor, J.F. et al. (2024). Intestinal organ chips for disease modelling and personalized medicine. *Nature Reviews Gastroenterology & Hepatology* 21 (11): 751–773.

28 Meglinski, I.V., Churmakov, D.Y., Bashkatov, A.N., et al. (2004). The enhancement of confocal images of tissues at bulk optical immersion. arXiv preprint physics/0401104.

29 Keyaerts, M., Caveliers, V., and Lahoutte, T. (2012). Bioluminescence imaging: looking beyond the light. *Trends in Molecular Medicine* 18 (3): 164–172.

30 Ozawa, T., Yoshimura, H., and Kim, S.B. (2013). Advances in fluorescence and bioluminescence imaging. *Analytical Chemistry* 85 (2): 590–609.

31 Shinde, A., Illath, K., Kasiviswanathan, U. et al. (2023). Recent advances of biosensor-integrated organ-on-a-chip technologies for diagnostics and therapeutics. *Analytical Chemistry* 95 (6): 3121–3146.

32 Gandelman, O., Allue, I., Bowers, K., and Cobbold, P. (1994). Cytoplasmic factors that affect the intensity and stability of bioluminescence from firefly luciferase in living mammalian cells. *Journal of Bioluminescence and Chemiluminescence* 9 (6): 363–371.

33 Dehghani, H., Guggenheim, J.A., Taylor, S.L. et al. (2018). Quantitative bioluminescence tomography using spectral derivative data. *Biomedical Optics Express* 9 (9): 4163–4174.

34 Yao, Z., Zhang, B.S., and Prescher, J.A. (2018). Advances in bioluminescence imaging: new probes from old recipes. *Current Opinion in Chemical Biology* 45: 148–156.

35 Lee, J. and Seliger, U.H.H. (1965). Absolute spectral sensitivity of phototubes and the application to the measurement of the absolute quantum yields of chemiluminescence and bioluminescence. *Photochemistry and Photobiology* 4 (6): 1015–1048.

36 Liu, J., Hanze, D., Huang, L. et al. (2024). AI-powered microfluidics: shaping the future of phenotypic drug discovery. *ACS Applied Materials & Interfaces* 16 (30): 38832–38851.

37 Bates, T.A., Gurmessa, S.K., Reyes-Weinstein, J.B., et al. (2024). An effective method of measuring nanobody binding kinetics and competition-based epitope mapping using biolayer interferometry. *bioRxiv*: 2024-10.

38 Weeramange, C.J., Fairlamb, M.S., Singh, D. et al. (2020). The strengths and limitations of using biolayer interferometry to monitor equilibrium titrations of biomolecules. *Protein Science* 29 (4): 1004–1020.

39 Berezin, M.Y. and Achilefu, S. (2010). Fluorescence lifetime measurements and biological imaging. *Chemical Reviews* 110 (5): 2641–2684.

40 He, Z., Wang, P., Liang, Y. et al. (2021). Clinically available optical imaging technologies in endoscopic lesion detection: current status and future perspective. *Journal of Healthcare Engineering* 2021 (1): 7594513.

41 Xiao, J., Ding, X., Huang, W. et al. (2024). Rotating machinery weak fault features enhancement via line-defect phononic crystal sensing. *Mechanical Systems and Signal Processing* 220: 111657.

42 Inoue, Y., Izawa, K., Kiryu, S. et al. (2008). Bioluminescent evaluation of the therapeutic effects of total body irradiation in a murine hematological malignancy model. *Experimental Hematology* 36 (12): 1634–1641.

43 Iafrate, M. and Fruhwirth, G.O. (2020). How non-invasive in vivo cell tracking supports the development and translation of cancer immunotherapies. *Frontiers in Physiology* 11: 154.

44 Henry, M. (2012). *in vivo Magnetic Resonance Imaging for Assessing the Integrity of the Blood-Tumour Barrier in a Mouse Model of Melanoma Brain Metastasis*. The University of Western Ontario (Canada).

45 Pirovano, G., Roberts, S., Kossatz, S., and Reiner, T. (2020). Optical imaging modalities: principles and applications in preclinical research and clinical settings. *Journal of Nuclear Medicine* 61 (10): 1419–1427.

46 Araújo-Gomes, N., Zambito, G., Johnbosco, C. et al. (2023). Bioluminescence imaging on-chip platforms for non-invasive high-content bioimaging. *Biosensors and Bioelectronics* 237: 115510.

47 Jiang, Y., Shi, X., Tang, C., and Wang, F. (2023). Beyond luciferase-luciferin system: modification, improved imaging and biomedical application. *Coordination Chemistry Reviews* 481: 215045.

48 Tahir, M. (2023). *BRET-Based Reporter Conjugated Luminescent Nanoparticles for Improved Bioluminescence Imaging*. University of California, Irvine.

49 Diao, W., Huang, X., Shen, L., and Zeng, Z. (2018). Diagnostic ability of blue laser imaging combined with magnifying endoscopy for early esophageal cancer. *Digestive and Liver Disease* 50 (10): 1035–1040.

50 O'farrell, A. C., S. D. Shnyder, G. Marston, P. L. Coletta, and J. H. Gill. (2013). Non-invasive molecular imaging for preclinical cancer therapeutic development. *British Journal of Pharmacology* 169 (4): 719–735.

51 Troy, T., Jekic-McMullen, D., Sambucetti, L., and Rice, B. (2004). Quantitative comparison of the sensitivity of detection of fluorescent and bioluminescent reporters in animal models. *Molecular Imaging* 3 (1): 9–23.

52 Virolainen, N. (2012). Antimicrobial detection illuminated: developing bioluminescent antibiotic biosensors based on bacterial gene regulatory elements.

53 Klerk, C.P.W., Overmeer, R.M., Niers, T.M.H. et al. (2007). Validity of bioluminescence measurements for noninvasive in vivo imaging of tumor load in small animals. *BioTechniques* 43 (sup 1): S7–S13.

54 Darne, C., Yujie, L., and Sevick-Muraca, E.M. (2013). Small animal fluorescence and bioluminescence tomography: a review of approaches, algorithms and technology update. *Physics in Medicine & Biology* 59 (1): R1–R64.

55 Chaudhury, S., Thakur, B., Chatterjee, S., and Ray, P. (2014). Molecular imaging aided improvement in drug discovery and development. *Current Biotechnology* 3 (3): 218–237.

56 Syed, A.J. and Anderson, J.C. (2021). Applications of bioluminescence in biotechnology and beyond. *Chemical Society Reviews* 50 (9): 5668–5705.

57 Close, D.M., Tingting, X., Sayler, G.S., and Ripp, S. (2011). in vivo bioluminescent imaging (BLI): noninvasive visualization and interrogation of biological processes in living animals. *Sensors* 11 (1): 180–206.

58 Polzer, W.E. (2019). Biodistribution studies of intratracheally applied nucleic acid-based drugs using multimodal optical tomography.

59 Sangeetha, B., Leroy, K.I., and Kumar, B.U. (2024). Harnessing bioluminescence: a comprehensive review of in vivo imaging for disease monitoring and therapeutic intervention. *Cell Biochemistry and Function* 42 (8): e70020.

60 Hong, Y., Yang, Y., Wang, C. et al. (2024). Luciferase-loaded calcium phosphate nanoparticles for persistent bioluminescence imaging of orthotopic breast tumors. *Analytical Chemistry* 96 (36): 14320–14325.

61 Jiang, T., Song, J., and Zhang, Y. (2023). Coelenterazine-type bioluminescence-induced optical probes for sensing and controlling biological processes. *International Journal of Molecular Sciences* 24 (6): 5074.

62 Huang, Z. and Li, M. (2024). Engineering of prevalent luciferase–luciferin pairs toward tunable light and applications in the field of bioluminescence imaging.

63 Sadikot, R.T. and Blackwell, T.S. (2005). Bioluminescence imaging. *Proceedings of the American Thoracic Society* 2 (6): 537–540.

64 Avci, P., Karimi, M., Sadasivam, M. et al. (2018). In-vivo monitoring of infectious diseases in living animals using bioluminescence imaging. *Virulence* 9 (1): 28–63.

65 O'Farrell, A.C. (2012). Development of in vivo tumour models for non-invasive proof-of-principle investigation of novel therapeutic agents. Engineering and characterisation of bioluminescent cell reporter systems for in vivo analysis of anti-cancer therapy pharmacodynamics. PhD diss., University of Bradford.

66 McCaffrey, A., Kay, M.A., and Contag, C.H. (2003). Advancing molecular therapies through in vivo bioluminescent imaging. *Molecular Imaging* 2 (2): 15353500200303124.

67 Appel, A.A., Anastasio, M.A., Larson, J.C., and Brey, E.M. (2013). Imaging challenges in biomaterials and tissue engineering. *Biomaterials* 34 (28): 6615–6630.

68 Balasubramanian, H., Hobson, C.M., Chew, T.-L., and Aaron, J.S. (2023). Imagining the future of optical microscopy: everything, everywhere, all at once. *Communications Biology* 6 (1): 1096.

69 O'Donoghue, S.I., Baldi, B.F., Clark, S.J. et al. (2018). Visualization of biomedical data. *Annual Review of Biomedical Data Science* 1 (1): 275–304.

70 Ma, X., Mao, M., He, J. et al. (2023). Nanoprobe-based molecular imaging for tumor stratification. *Chemical Society Reviews* 52: 6447–6496.

71 Moliner, D., Fabio, F.N.-B., and Vendrell, M. (2024). Recent advances in minimal fluorescent probes for optical imaging. *Current Opinion in Chemical Biology* 80: 102458.

72 Greig, J.C., Tipping, W.J., Graham, D. et al. (2024). New insights into lipid and fatty acid metabolism from Raman spectroscopy. *Analyst* 149 (19): 4789–4810.

73 Desai, D., Grace, W., and Zaman, M.H. (2011). Tackling HIV through robust diagnostics in the developing world: current status and future opportunities. *Lab on a Chip* 11 (2): 194–211.

74 Wang, Y., Cao, D., Chen, S.-L. et al. (2021). Current trends in three-dimensional visualization and real-time navigation as well as robot-assisted technologies in hepatobiliary surgery. *World Journal of Gastrointestinal Surgery* 13 (9): 904–922.

75 Tian, T., Yang, Z., and Li, X. (2021). Tissue clearing technique: recent progress and biomedical applications. *Journal of Anatomy* 238 (2): 489–507.

76 Ale, A., Ermolayev, V., Herzog, E. et al. (2012). FMT-XCT: in vivo animal studies with hybrid fluorescence molecular tomography–X-ray computed tomography. *Nature Methods* 9 (6): 615–620.

77 An, Y., Wang, K., and Tian, J. (2018). Recent methodology advances in fluorescence molecular tomography. *Visual Computing for Industry, Biomedicine, and Art* 1: 1–11.

78 Wu, M. and Shu, J. (2018). Multimodal molecular imaging: current status and future directions. *Contrast Media & Molecular Imaging* 2018 (1): 1–12.

79 Raza, A., Zulfiqar, H., Gong, Z., et al. (2024). A comprehensive review on biomedical sensors: technological advancements, applications in molecular informatics, and future trends.

80 Bai, J.-W., Qiu, S.-Q., and Zhang, G.-J. (2023). Molecular and functional imaging in cancer-targeted therapy: current applications and future directions. *Signal Transduction and Targeted Therapy* 8 (1): 89.

81 Icha, J., Weber, M., Waters, J.C., and Norden, C. (2017). Phototoxicity in live fluorescence microscopy, and how to avoid it. *BioEssays* 39 (8): 1700003.

82 Cox, G. (ed.) (2019). *Fundamentals of Fluorescence Imaging*. CRC Press.

83 Royer, C.A. and Scarlata, S.F. (2008). Fluorescence approaches to quantifying biomolecular interactions. *Methods in Enzymology* 450: 79–106.

84 Monici, M. (2005). Cell and tissue autofluorescence research and diagnostic applications. *Biotechnology Annual Review* 11: 227–256.

85 Mancini, T., Mosetti, R., Marcelli, A. et al. (2022). Terahertz spectroscopic analysis in protein dynamics: Current status. *Radiation* 2 (1): 100–123.

86 Brady, D.J. (2009). *Optical Imaging and Spectroscopy*. Wiley.

87 Yu, B., Ferris, D.G., Liu, Y., and Nagarajan, V.K. (2014). Emerging optical techniques for detection of oral, cervical and anal cancer in low-resource settings. *Austin Journal of Biomedical Engineering* 1 (2): 1007.

88 van der Vorst, J.R., Schaafsma, B.E., Hutteman, M. et al. (2013). Near-infrared fluorescence-guided resection of colorectal liver metastases. *Cancer* 119 (18): 3411–3418.

89 Cassinotti, E. (2017). Clinical application of fluorescence guided surgery.

90 Pinheiro, L.G.P., Vasques, P.H.D., Maia, M. et al. (2011). Sentinel lymph node biopsy: actual topics. In: *Breast Cancer-Recent Advances in Biology, Imaging and Therapeutics* (ed. S. Done), 397–426. IntechOpen.

91 Bu, L., Shen, B., and Cheng, Z. (2014). Fluorescent imaging of cancerous tissues for targeted surgery. *Advanced Drug Delivery Reviews* 76: 21–38.

92 Beyer, T., Bidaut, L., Dickson, J. et al. (2020). What scans we will read: imaging instrumentation trends in clinical oncology. *Cancer Imaging* 20: 1–38.

93 Zhang, P., Fan, G., Xing, T. et al. (2021). UHR-DeepFMT: ultra-high spatial resolution reconstruction of fluorescence molecular tomography based on 3-D fusion dual-sampling deep neural network. *IEEE Transactions on Medical Imaging* 40 (11): 3217–3228.

94 Wang, Z., Chen, Z., Zhang, Z. et al. (2024). High-orderly assembled organic fluorescent materials for high-resolution bio-sensing: a review. *Biomaterials Science* 12: 2019–2032.

95 Fang, H., Chen, Y., Jiang, Z. et al. (2023). Fluorescent probes for biological species and microenvironments: from rational design to bioimaging applications. *Accounts of Chemical Research* 56 (3): 258–269.

96 Delgado-Rodriguez, P., Brooks, C.J., Vaquero, J.J., and Munoz-Barrutia, A. (2022). Innovations in ex vivo light sheet fluorescence microscopy. *Progress in Biophysics and Molecular Biology* 168: 37–51.

97 James, M.L. and Gambhir, S.S. (2012). A molecular imaging primer: modalities, imaging agents, and applications. *Physiological Reviews* 92 (2): 897–965.

98 Choy, G., Choyke, P., and Libutti, S.K. (2003). Current advances in molecular imaging: noninvasive in vivo bioluminescent and fluorescent optical imaging in cancer research. *Molecular Imaging* 2 (4): 303–312.

99 Rego, G.N.A., Nucci, M.P., Mamani, J.B. et al. (2020). Therapeutic efficiency of multiple applications of magnetic hyperthermia technique in glioblastoma using aminosilane coated iron oxide nanoparticles: in vitro and in vivo study. *International Journal of Molecular Sciences* 21 (3): 958.

100 Wang, H., Xiaoyu, M., Yang, J. et al. (2019). Brain imaging with near-infrared fluorophores. *Coordination Chemistry Reviews* 380: 550–571.

101 Klose, A.D. and Paragas, N. (2018). Automated quantification of bioluminescence images. *Nature Communications* 9 (1): 4262.

102 Paley, M.A. and Prescher, J.A. (2014). Bioluminescence: a versatile technique for imaging cellular and molecular features. *MedChemComm* 5 (3): 255–267.

103 Ntziachristos, V. (2006). Fluorescence molecular imaging. *Annual Review of Biomedical Engineering* 8 (1): 1–33.

104 Capsi-Morales, P., Piazza, C., Grioli, G. et al. (2023). The SoftHand Pro platform: a flexible prosthesis with a user-centered approach. *Journal of Neuroengineering and Rehabilitation* 20 (1): 20.

105 Yeh, H.-W. and Ai, H.-W. (2019). Development and applications of bioluminescent and chemiluminescent reporters and biosensors. *Annual Review of Analytical Chemistry* 12 (1): 129–150.

106 Roda, A., Guardigli, M., Michelini, E., and Mirasoli, M. (2009). Bioluminescence in analytical chemistry and in vivo imaging. *TrAC Trends in Analytical Chemistry* 28 (3): 307–322.

107 Tiwari, D.K., Tiwari, M., and Jin, T. (2020). Near-infrared fluorescent protein and bioluminescence-based probes for high-resolution in vivo optical imaging. *Materials Advances* 1 (5): 967–987.

108 Dereure, E. (2024). Quantitative analysis of bioluminescent signals in preclinical imaging. PhD diss., Sorbonne Université.

109 Sargazi, S., Fatima, I., Kiani, M.H. et al. (2022). Fluorescent-based nanosensors for selective detection of a wide range of biological macromolecules: a comprehensive review. *International Journal of Biological Macromolecules* 206: 115–147.

110 Licha, K. and Olbrich, C. (2005). Optical imaging in drug discovery and diagnostic applications. *Advanced Drug Delivery Reviews* 57 (8): 1087–1108.

111 Kenry, Y.D. and Liu, B. (2018). Recent advances of optical imaging in the second near-infrared window. *Advanced Materials* 30 (47): 1802394.

112 Cabello, M.C., Chen, G., Melville, M.J. et al. (2024). Ex tenebris lux: illuminating reactive oxygen and nitrogen species with small molecule probes. *Chemical Reviews* 124 (16): 9225–9375.

113 Jenkins, D.E., Oei, Y., Hornig, Y.S. et al. (2003). Bioluminescent imaging (BLI) to improve and refine traditional murine models of tumor growth and metastasis. *Clinical & Experimental Metastasis* 20: 733–744.

114 Campbell, J.M., Gosnell, M., Agha, A. et al. (2024). Label-free assessment of key biological autofluorophores: material characteristics and opportunities for clinical applications. *Advanced Materials* 36 (42): 2403761.

7

Innovative Diagnostic Imaging Techniques and Protocols

Kamini Lamba, Shalli Rani, Ayush Dogra, and Ankita Sharma

Chitkara Institute of Engineering and Technology, Chitkara University, Punjab, India

7.1 Introduction

Medical imaging has brought great revolution in diagnosing disease, planning, and monitoring treatment. Although conventional imaging modalities such as magnetic resonance imaging (MRI), computed tomography (CT), positron emission tomography (PET), ultrasound, and X-ray are widely used by healthcare professionals for making decisions, these methods often result in various challenges such as variation in image resolution, contrast enhancement, radiation exposure, and computational efficiency. In contrast, cutting-edge imaging approaches have been developed to overcome the existing challenges which include integrating advanced approaches of artificial intelligence (AI), molecular imaging, functional imaging, and multi-modal imaging in the medical field for improvement in the diagnostic accuracy, minimizing invasive methods, and facilitating efficient outcomes [1]. Figure 7.1 illustrates spectrum of imaging techniques whose description has been given below.

7.1.1 Evolution of Multi-Modal and Hybrid Imaging

The process of integrating distinct imaging methods for enhancing diagnostic capabilities has brought great transformation in the medical field. For instance, integrating MRI or CT with PET may result in combining their potential of anatomical precision and functional imaging to have in-depth insights into the biology of tumor, neurological disorders, and cardiovascular diseases [2]. In addition, Diffusion Tensor Imaging (DTI) and Functional Magnetic Resonance Imaging (fMRI) also resulted in enhanced neuroimaging based on the mapping of brain connectivity and identification of abnormal growth of tissues in neurological disorders namely Alzheimer, Parkinson, and brain tumor [3].

Biomedical Imaging Technology: Signal Processing Strategies and Innovations, First Edition.
Edited by Ayush Dogra, Shalli Rani, and Ankita Sharma.

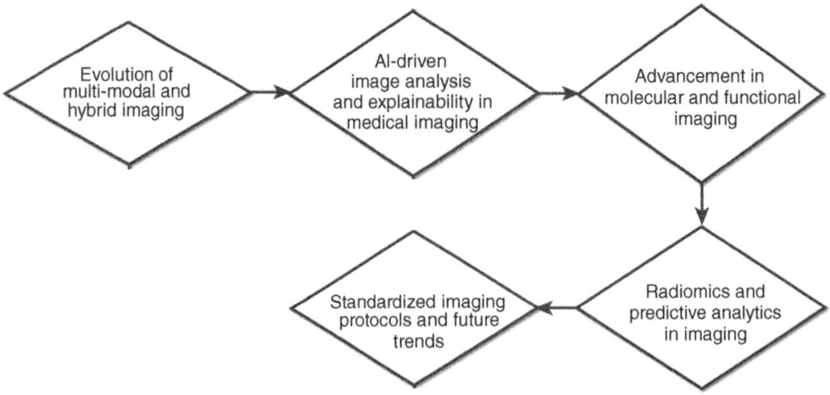

Figure 7.1 Spectrum of innovative imaging techniques.

7.1.2 AI-Driven Image Analysis and Explainability in Medical Imaging

Emergence and development of deep learning and explainable AI resulted in improved and enhanced interpretation of image analysis. It also helped radiologists in making accurate decision with higher precision while diagnosing presence of abnormal tissues. Thus, integrating AI with imaging protocols helps in performing automated segmentation, classification, and detection of abnormalities. Utilizing convolutional neural networks, generative adversarial networks, and transformer models can also result in enhancement of image reconstruction, denoising especially when it comes to low image contrast [4]. Transparency of the achieved predicted results can be visualized via integrating advanced techniques of artificial intelligence such as LIME, SHAP, Grad-CAM, and IG which ultimately results in gaining trustworthiness among experts to achieve reliable outcomes [5].

7.1.3 Advancement in Molecular and Functional Imaging

To enable non-invasive tracking of cellular and molecular processes, facilitate early detection of abnormal tissues and planning of treatment, a lot of innovations are done in molecular imaging and theranostics. Moreover, exploring optical imaging, Raman spectroscopy, and fluorescence-guided imaging can help in diagnosing a particular disease at the cellular level, whereas hyperpolarized MRI and metabolic imaging techniques help in assessing metabolism of tumor while focusing on its treatment [6].

7.1.4 Radiomics and Predictive Analytics in Imaging

Radiomics has been emerged as a powerful tool in precision medicine which can extract high-dimensional imaging features. Integrating radiomics with machine learning can help in predicting progress of disease, responding its treatment and prognosis of patients based on the analysis of texture, shape, intensity features from medical images. As a result, it can help in providing valuable biomarker for classifying tumor [7].

7.1.5 Standardized Imaging Protocols and Future Trends

Standardizing imaging protocols has become significant in terms of reproducibility and clinical applicability with the increase in big data. There are various organizations, namely Quantitative Imaging Biomarkers Alliance (QIBA) and Radiological Society of North America (RSNA) are strictly following guidelines to harmonize imaging workflows while focusing to improve consistency across distinct scanners and institutions. However, real-time AI-based imaging diagnostics, cloud-based imaging platforms, non-invasive brain stimulation methods when integrated with one another can demonstrate great potential in personalized medicine and precision healthcare [8].

Thus, innovative diagnostic imaging techniques and protocols hold promise in reshaping the process of diagnosing disease in the medical field based on the integration of the AI, multi-modal imaging, molecular imaging, and radiomics. As a result, these methods can improve accuracy of diagnostic process, minimize false outcomes, and enhance quality of patients.

7.2 Diagnosing Imaging Methods

To diagnose brain tumor efficiently, there have been continuous advancements in the diagnostic imaging techniques. Conventional methods such as cerebral angiography and X-ray often resulted in less contrast of soft tissues. To overcome the drawbacks of these techniques, MRI, CT, and PET were introduced to provide more fine detailed images to aid healthcare professionals in their decision-making process in the healthcare. Thereafter, deploying machine learning, deep learning, and advanced AI also resulted in predictions based on the features extracted from the medical input data, but their black box nature also resulted in lack of transparency and in-depth insights into achieved outcomes. To overcome these challenges and gain trustworthiness among experts in the healthcare, advanced approaches of AI are also introduced in the healthcare to assist radiologist. Thus, the chapter highlights the conventional and advanced innovative diagnostic imaging methods to detect disease along with their potential and drawbacks.

7.2.1 Conventional Diagnostic Imaging Techniques

7.2.1.1 X-Ray Radiography and Its Limitations

It was among the first imaging approaches used to detect abnormality lying within skull associated with brain tumor during the early 20th century. It usually passes high-energy electromagnetic radiation via skull where distinct densities from multiple tissues are absorbed at varied rates and produces two-dimensional projection image of the brain. Based on the property of absorbing more radiation than soft tissues, X-rays are specifically used to identify skull fractures and abnormal masses of brain tumor; however, it also represents its limited ability of differentiating structures comprising soft tissues [9].

7.2.1.2 Pneumoencephalography (PEG): A Historical Perspective and Its Limitations

It was also developed in the early 20th century to detect brain tumor. The process includes replacement of cerebrospinal fluid (CSF) from the ventricular system with air or another gas for enhancing contrast of brain structure in x-ray images, though the procedure often resulted in headaches, nausea, vomiting because of rapid decrease in the pressure of CSF. These highly invasive, painful, severe complications led to its discontinuity after the advent of CT and MRI [10].

7.2.1.3 Cerebral Angiography: Detecting Tumor-Related Vascular Abnormalities

To have appropriate visualization among the blood vessels of brain, a technique utilizing contrast agents and x-ray was introduced and given name of cerebral angiography [11]. It provides comprehensive imaging of arterial and venous systems in the brain due to its ability of bringing variation in the surrounding blood vessels. As a result, it can easily detect tumor-related vascular abnormalities. In addition, there are distinct kinds of cerebral angiography such as conventional angiography which produces high-resolution images of brain's vascular network based on the procedure of catheter for injecting contrast dye into the arterial system of brain. Digital subtraction angiography subtracts bone and other structures from the final image and results in displaying only blood vessels which enhances contrast resolution and gives clear vision of blood vessels. Three-dimensional angiography generates complete picture of anomalies relevant to tumor and assist in planning of pre-surgical procedures. Although these techniques have various advantages, invasiveness, radiation exposure, and limited soft tissue resolution are a few drawbacks which are overcome by introducing advanced imaging techniques.

7.2.2 Advanced Imaging Modalities in Brain Tumor Detection

Following is the description of advanced imaging modalities in detection of brain tumor.

7.2.2.1 Computed Tomography (CT) and Its Advancements

This technique was introduced in the 1970s to revolutionize detection of brain tumor. As a result, it offered high-resolution cross-sectional imaging and improved visibility of abnormal growth of cells lying within brain. This approach has three distinct kinds out of which multi-detector CT scan helped in enabling rapid volumetric imaging, spectral CT, and photon-counting computed tomography resulted in enhancement of tumor differentiation based on the analyses of tissue-specific x-ray absorption [12].

7.2.2.2 Magnetic Resonance Imaging (MRI) and Functional Variants

This technique [13] is considered as a gold standard when it comes to brain tumor imaging due to its ability of providing excellent soft-tissue contrast. It also supports variants like diffusion-weighted imaging and DTI responsible for assessing microstructure of malignancy and infiltered patterns. In addition, functional MRI helps in mapping of brain activity and assists in planning of pre-surgical procedures. Moreover, different types of malignancies are differentiated by magnetic resonance spectroscopy based on the metabolic signatures.

7.2.2.3 Positron Emission Tomography (PET) and Hybrid Imaging

Positron Emission Tomography (PET): This technique [14] is mostly preferred in the fields of oncology, neurology, and cardiology to detect presence of disease such as brain tumor, Alzheimer, and cardiovascular and monitor its progress for planning appropriate treatment. The process utilizes radiotracers such as positron-emitting isotopes namely Fluorine-18 for providing metabolic as well as functional details about tissues and organs by injecting tracer into the body for proper accumulation through the target tissues. Thereafter, PET scanners detect produced gamma rays due to the interaction of emitted positrons with nearby electrons [15]. In addition, the most commonly utilized PET tracer, 18F-fluorodeoxyglucose (FDG) is having similarity with glucose due to which increased metabolic activity causes cancerous tissues to absorb more FDG. This process further increases radiotracer uptake and makes identification of tumor easy [16]. As a result, this technique helps in providing functional imaging instead of structural data and has ability to assess treatment response in cancer therapy while focusing on early-stage detection of disease. However, it is a very expensive method and requires cyclotron for producing tracers. Moreover, it also causes less spatial resolution when compared to the MRI and CT.

Hybrid Imaging: PET–CT and PET–MRI Integrating PET with other imaging modalities of CT and MRI may result in enhancement of anatomical localization and diagnostic accuracy. PET–CT results in integration of metabolic and anatomical information. The hybrid technique of PET–CT [17] results in improved localization of lesion, quicker processing time than individual PET and CT scans, increased diagnostic outcomes in oncology. However, it is an expensive technology and may cause higher radiation dose. On the other hand, the hybrid technique of PET–MRI [18] results in less radiation dose when compared to the PET–CT and has ability to process functional and structural data simultaneously. In addition, it also has superiority among soft-tissue contrast, specifically in the brain and musculoskeletal structure in neurology. However, it takes more time in scanning the input data and also involves higher operational cost with limited availability of the required resources.

7.3 Comparison of Innovative Diagnostic Imaging Techniques and Protocols

Figure 7.2 shows innovative diagnostic techniques used in the healthcare for detection of disease, whereas Table 7.1 represents the comparison of innovative diagnostic imaging techniques and protocols based on their modality, keyfeatures, advantages, challenges, and example applications.

The chapter has highlighted the strengths and drawbacks of the various methods used for the detection of brain tumor. Although these deployed methods such as deep-learning-based MRI and cutting-edge offer accurate and improved outcomes, interpretability and real-time clinical integration are still major concerns in the medical field. The traditional protocols like contrast-enhanced magnetic resonance imaging and spectroscopy are considered as gold standards because of their reliable property, but lack in automation and interpretation are their major drawbacks. Therefore, approaches of hybrid AI have been deployed to provide more accurate, interpretable, and efficient outcomes in diagnosing disease. However, biasness in dataset, high cost, and resistance from the experts can become major challenges which must be addressed while integrating in the clinical settings for improved patient outcomes.

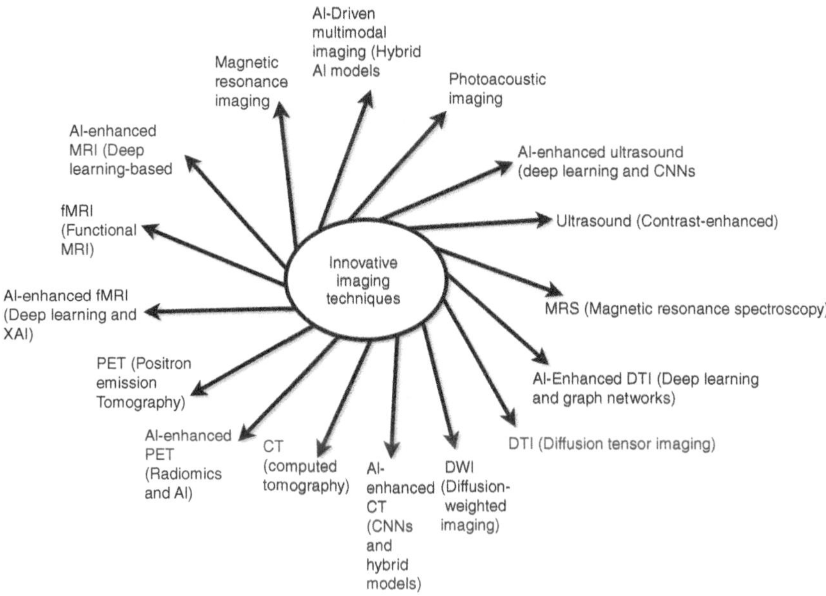

Figure 7.2 Illustration of innovative imaging techniques.

7.4 Challenges in Innovative Diagnostic Imaging

Though there have been advancements in the medical field while developing an automated system, following is the description of various challenges of innovative diagnostic imaging as shown in Figure 7.3.

7.4.1 Data Heterogeneity and Standardization Issues

Data heterogeneity is one of the major issues in the diagnosing imaging. As the input dataset is usually acquired from different sources it can belong to different formats thus causing variation in contrast, resolutions, and protocols [19]. These variations can also be the result of shifting in domains which further lead to inconsistencies in the performance of model while deploying them in the clinical settings [20]. In addition, maintaining standardization of dataset has become very crucial for good generalization among unseen and distinct datasets [21]. Thus, both these factors need to be addressed while diagnosing disease in the healthcare.

7.4.2 High Computational and Infrastructural Cost

Deploying automated models, specifically deep learning, requires need to handle large-scale medical imaging dataset for providing appropriate training to achieve

Table 7.1 Comparison of innovative diagnostic imaging techniques and protocols.

Technique	Modality	Key features	Advantages	Challenges	Example applications
Magnetic resonance imaging	Non-ionizing	High-resolution soft tissue contrast, multiparametric imaging	No radiation exposure, detailed brain tissue analysis	Expensive, long scanning time, motion artifacts	Brain tumor localization, structural analysis
AI-Enhanced MRI (Deep Learning-based)	Non-ionizing	Automated tumor segmentation and classification using CNNs, GANs	Faster, more accurate, reduces radiologist workload	Requires large labeled datasets, potential bias	Automated glioma grading, tumor boundary detection
fMRI (Functional MRI)	Non-ionizing	Measures brain activity via blood oxygenation (BOLD signal)	Non-invasive, detects functional abnormalities	Low temporal resolution, motion sensitivity	Pre-surgical mapping, neurodegenerative diseases
AI-Enhanced fMRI (Deep Learning and XAI)	Non-ionizing	Identifies hidden biomarkers using explainable AI models	Improved interpretability, early disease detection	High computational cost, complex validation	Functional brain disorder diagnosis, tumor prognosis
PET (Positron Emission Tomography)	Ionizing	Uses radiotracers to assess metabolic activity	Detects early-stage tumors, functional imaging	Radiation exposure, costly radiotracers	Tumor grading, metabolic profiling
AI-Enhanced PET (Radiomics and AI)	Ionizing	Extracts quantitative features from PET images for AI-based diagnosis	Early-stage tumor identification, improved precision	High cost, requires standardization	Tumor radiomics analysis, personalized treatment
CT (Computed Tomography)	Ionizing	3D reconstruction, fast imaging	Rapid scan, effective for emergency cases	High radiation dose, less contrast than MRI	Tumor detection, emergency diagnosis
AI-Enhanced CT (CNNs and Hybrid Models)	Ionizing	AI-driven segmentation and anomaly detection	Reduces manual errors, improves sensitivity	Requires extensive validation, potential false positives	AI-driven tumor detection, automated triaging

DWI (Diffusion-Weighted Imaging)	Non-ionizing	Measures water diffusion in tissues	High sensitivity for tumor detection, no contrast needed	Susceptible to motion artifacts	Early-stage glioma detection, stroke assessment
DTI (Diffusion Tensor Imaging)	Non-ionizing	Maps white matter tracts in the brain	Useful for surgical planning, connectivity analysis	Complex interpretation, long acquisition time	Brain tumor invasion analysis, neuroconnectivity
AI-Enhanced DTI (Deep Learning and Graph Networks)	Non-ionizing	Detects microstructural abnormalities in white matter	Advanced connectivity mapping, personalized analysis	Requires robust validation, complex computations	AI-assisted neuro-surgical planning, tumor spread analysis
MRS (Magnetic Resonance Spectroscopy)	Non-ionizing	Analyzes biochemical composition of tissues	Differentiates tumor types, metabolic insights	Low spatial resolution, long scan time	Tumor grading, glioma characterization
Ultrasound (Contrast-Enhanced)	Non-ionizing	Uses microbubbles for improved contrast	Real-time imaging, portable	Limited penetration, operator-dependent results	Intraoperative tumor visualization
AI-Enhanced Ultrasound (Deep Learning and CNNs)	Non-ionizing	AI-assisted tumor detection and segmentation	Improves accuracy in real-time imaging	Operator training required, data quality dependency	AI-assisted real-time tumor detection
Photoacoustic Imaging	Hybrid	Combines optical and ultrasound imaging	High contrast and depth, functional imaging	Limited clinical adoption, expensive	Vascular tumor assessment, functional imaging
AI-Driven Multimodal Imaging (Hybrid AI models)	Multi-modal	Integrates MRI, PET, CT, and histopathology using deep learning	Higher diagnostic accuracy, robust feature fusion	Requires high computational power, data integration complexity	AI-powered tumor diagnosis, personalized treatment planning

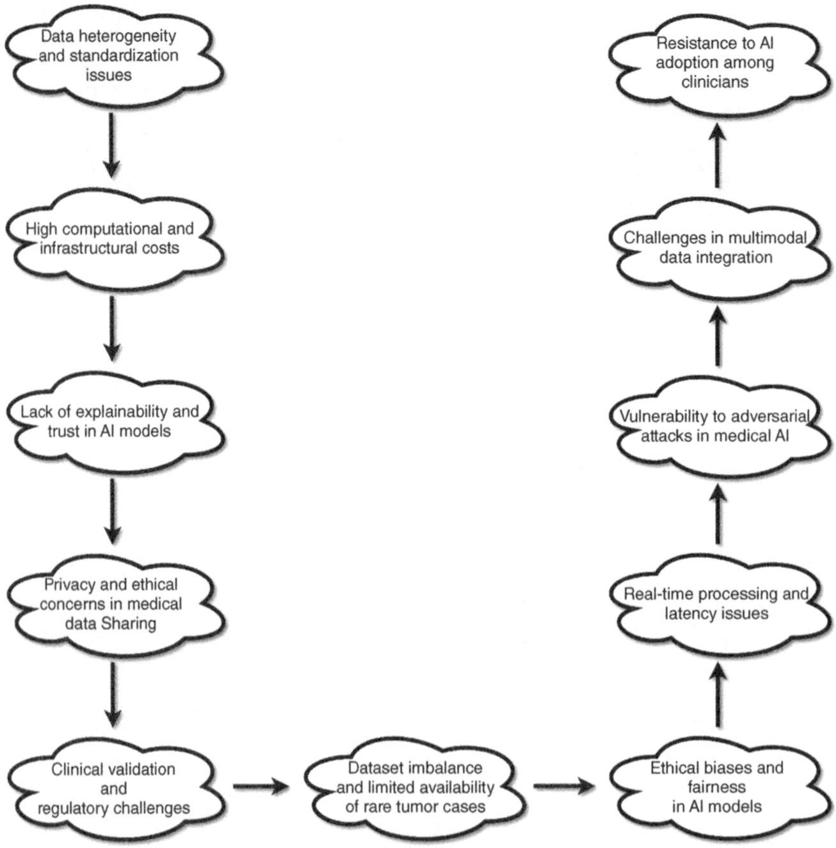

Figure 7.3 Challenges of innovative imaging techniques.

efficient results while diagnosing presence of abnormal tissues. However, these techniques require high-performance graphical processing units, and cloud-based computational resources which can again develop challenges and result in resource constraints [22]. In addition, implementing the advanced techniques of AI in real-time applications can also demand for increased memory and processing speed which can again raise a concern due to the limited computational infrastructure [23]. Thus, high computational and infrastructural cost become great barriers which need to be addressed while diagnosing disease in the medical field.

7.4.3 Lack of Explainability and Trust in Artificial-Intelligence Models

Due to the properties of analyzing medical images efficiently and accurately, many researchers suggested a framework comprising deep learning models to detect brain tumor, but the black box nature of such models raised a concern [24] regarding reasoning behind their predictions as it becomes very difficult for the experts in understanding the cause responsible for achieving that prediction. Thus, advanced methods such as explainable AI are suggested to be deployed in the automated system for enhancing transparency and gain trust among experts [25]. However, these methods may have challenges such as lacking of standardization and clinical validation.

7.4.4 Privacy and Ethical Concerns in Medical Data Sharing

Accessing large-scale datasets while focusing on its diverse form has become one of the main requirements to train the model by AI-based diagnostic methods. However, it may also result in restricting the data of patients to be shared among one another due to their privacy concerns [26]. As a result, majority of the hospitals and research institutes experience such challenges to preserve the data of patients while following ethical concerns. To have security among the shared data, researchers provided framework integrating federated learning with blockchain based decentralized AI model and resulted in enabling collaborative method to train the system without any need of direct exchange of data [27]. But implementing the suggested framework in compliance with the real-time application is still a major concern as it may require significant advancements in the form of its encryption procedure, privacy in differential aspect, and securing multi-party computation.

7.4.5 Clinical Validation and Regulatory Challenges

Despite the various advancements done in the AI-based automated diagnostic methods, it is still crucial to implement such methods in the clinical settings for their appropriate validation. This can be the result of lack of generalization due to diverse patients data which lead toward errors along with biasness issues [28]. In addition, experts must be trained enough for utilizing these methods in the real-time clinical applications which can again raise an issue as it may cause burden for them too due to an individual expertise.

7.4.6 Dataset Imbalance and Limited Availability of Rare Tumor Cases

Most of the datasets available publicly contain common types of tumor where rare tumor kinds are very limited. It causes issues of data imbalance which leads the developed model to have more biasness toward majority of the class and causes poor generalization [29]. In addition, it is also a great challenge to obtain high-quality annotated medical data as it requires authenticity of experts to have appropriate curation of data that further results in time-consuming process [30].

7.4.7 Ethical Biases and Fairness in AI Models

Ethical biases and fairness in AI-based automated models are again a challenge as these models can inherit biases from the trained data which can lead toward generation of disparity in the process of diagnosing disease when distinct demographic groups are considered [28]. Moreover, various factors such as age, gender, socioeconomic background, and ethnicity can cause variation in the MRI data. The impact caused by such variations needs to be addressed in the future work for appropriate utilization of such AI models [31].

7.4.8 Real-Time Processing and Latency Issues

Deployed automated models such as three-dimensional convolutional neural network and transformer-based models often cause high latency [32] due to the requirement of substantial computational power. As a result, there are various hospitals which may have less GPU resources and less up-to-date imaging methods may not have enough features to cooperate with the time-sensitive nature of brain tumor imaging. Thus, it causes issues in real-time processing and latency [33].

7.4.9 Vulnerability to Adversarial Attacks in Medical AI

Due to the susceptible nature of adversarial attacks in the AI-based models, it can lead the proposed AI model to generate false predictions based on MRI [34]. Thus, attackers can misguide or manipulate the actual outcomes and causes false treatment to the patients based on the incorrect decision [35]. Thus, it is necessary to have defense procedure against such attackers to make the model secure which will result in reliable method in the medical field.

Other than these challenges, integrating multimodal data [36, 37] and resistance to AI [38] adoption in clinical settings are still great barriers which need to be addressed in the future work to diagnose presence of disease in the medical field.

7.5 Future Directions in Innovative Diagnostic Imaging for Brain Tumor Detection

Based on the advancements in the diagnostic imaging methods used for diagnosing disease in the medical field, various approaches need to integrate in the future to enhance the reliability among the automated systems. These techniques include explainable AI, multi-modal imaging, data fusion, low-cost and portable imaging solutions, AI and quantum computing, non-invasive tumor monitoring, and early detection in the medical field as shown in Figure 7.4.

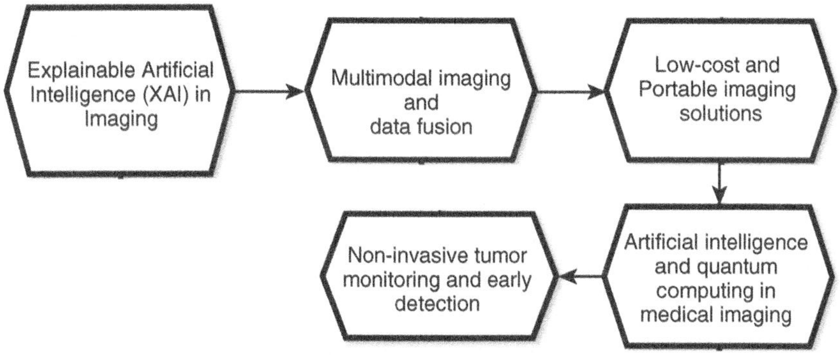

Figure 7.4 Future directions of innovative imaging techniques.

7.5.1 Explainable AI (XAI) in Imaging

Deploying traditional deep learning methods in the medical field results in generating predictions based on the extracted features and provide accurate results based on their properties of analyzing medical images given as an input but the black-box nature of these models usually considered as a challenge for the experts due to the lack of the in-depth sights behind achieving the predictions. Therefore, it is very crucial to integrate explainable AI [39] such as local interpretable model-agnostic explanations, Shapley additive explanations, gradient-weighted class activation mapping, integrated gradients in the future models for enhancing reliability and gain transparency among the predictions made by these traditional methods in future. Thus, the suggested framework can also result in improved and accurate diagnostic method for diagnosing disease in the healthcare.

7.5.2 Multimodal Imaging and Data Fusion

For enhancing tumor characterization in the automated methods, various modalities such as MRI, CT, PET, and histopathology are integrated with one another while developing multi-modal fusion-based deep learning model. It resulted in improved grading of tumor, segmented accuracy, excellent classification as compared to the single-modality methods [40]. However, these methods often result in standardization of data, storage related requirements along with complexity among computational resources. To overcome such challenges, federated learning and cloud-based artificial methods can be integrated in the future while developing automated system which will lead in enabling privacy in the shared data while performing collaborative research.

7.5.3 Low-Cost and Portable Imaging Solutions

It is crucial to access high-quality diagnostic imaging in low-resource settings as it becomes almost impractical as well as expensive to utilize MRI and PET scanners. Advancements in these low-field MRI in the form of portable scanners resulted in demonstrating great potential in diagnosing brain tumor while minimizing requirements of infrastructure [41]. Thus, integrating edge-based AI-driven methods and mobile health techniques will result in remote detection to offer more accessibility in the telemedicine field.

7.5.4 AI and Quantum Computing in Medical Imaging

Due to the potential of revolutionizing image reconstruction, quantum computing-based approach has been used in the medical field which also results in denoising and real-time processing of complicated medical data. These techniques also include variational quantum classifiers and quantum enhanced neural networks which have great potential in accelerating pipelines of image processing. Thus, it results in improved segmented tumor [42]. In future, quantum techniques can be integrated with AI to enable detection of tumor in instantaneous form.

7.5.5 Non-Invasive Tumor Monitoring and Early Detection

It has become crucial to detect brain tumor at an early stage as the deployed imaging methods have ability to detect abnormality at an advanced stage. Thus, AI-based biomarkers and non-invasive brain stimulation [43] based procedures can be used in the future while developing an automated method to diagnose presence of disease as these methods have ability to provide an individual brain neural activity which can be integrated with other techniques to achieve more accurate, transparent, and efficient results in neuro-oncology field.

7.6 Conclusion

The evolution of diagnostic imaging techniques and protocols has significantly enhanced the accuracy and reliability of brain tumor detection, classification, and treatment planning. From early conventional methods such as X-ray radiography, pneumoencephalography, and cerebral angiography to modern multi-modal MRI, PET, and AI-driven imaging, the field has undergone remarkable advancements. These innovations have improved tumor localization, characterization, and real-time surgical guidance, ultimately leading to better patient outcomes. The integration of AI and Explainable AI (XAI) has further revolutionized brain tumor diagnosis by enabling automated detection, segmentation, and classification, while enhancing interpretability for clinicians. Advanced imaging protocols, including radiomics, functional MRI, and intraoperative imaging, are continuously evolving to provide precise and personalized diagnostic insights. However, challenges such as high costs, accessibility issues, and standardization of AI models remain key barriers to widespread implementation. Looking ahead, future developments in quantum imaging, AI-augmented diagnostics, and portable imaging systems are expected to further improve early tumor detection and real-time monitoring. The integration of multimodal imaging, deep learning, and advanced biomarkers will continue to shape the future of precision medicine in neuro-oncology, fostering non-invasive, highly accurate, and patient-centered diagnostic solutions.

References

1 Palmer, G.M., Boruta, R.J., Viglianti, B.L. et al. (2010). Non-invasive monitoring of intra-tumor drug concentration and therapeutic response using optical spectroscopy. *Journal of Controlled Release* 142 (3): 457–464.

2 Quick, H.H. (2015). PET/MRI: physics, instrumentation, and systems. In: *Molecular and Multimodality Imaging in Cardiovascular Disease* (ed. T.H. Schindler, R.T. George, and J.A.C. Lima), 1–11. Cham: Springer International Publishing.

3 Shaffer, A., Kwok, S.S., Naik, A. et al. (2022). Ultra-high-field MRI in the diagnosis and management of gliomas: a systematic review. *Frontiers in Neurology* 13: 857825.

4 Ardila, D., Kiraly, A.P., Bharadwaj, S. et al. (2019). End-to-end lung cancer screening with three-dimensional deep learning on low-dose chest computed tomography. *Nature Medicine* 25 (6): 954–961.

5 Holzinger, A., Langs, G., Denk, H. et al. (2019). Causability and explainability of artificial intelligence in medicine. *Wiley Interdisciplinary Reviews: Data Mining and Knowledge Discovery* 9 (4): e1312.

6 Lambin, P., Rios-Velazquez, E., Leijenaar, R. et al. (2012). Radiomics: extracting more information from medical images using advanced feature analysis. *European Journal of Cancer* 48 (4): 441–446.

7 Ferda, J., Ferdov'a, E., Hes, O. et al. (2017). PET/MRI: multiparametric imaging of brain tumors. *European Journal of Radiology* 94: A14–A25.

8 Bergmann, T.O., Karabanov, A., Hartwigsen, G. et al. (2016). Combining non-invasive transcranial brain stimulation with neuroimaging and electrophysiology: current approaches and future perspectives. *NeuroImage* 140: 4–19.

9 Castillo, M. (2014). History and evolution of brain tumor imaging: insights through radiology. *Radiology* 273 (2S): S111–S125.

10 Ishaque, M., Wallace, D.J., and Grandhi, R. (2017). Pneumoencephalography in the workup of neuropsychiatric illnesses: a historical perspective. *Neurosurgical Focus* 43 (3): E10.

11 Culbreth, G.G., Walker, A.E., and Curry, R.W. (1950). Cerebral angiography in "brain tumor suspects". *Journal of Neurosurgery* 7 (2): 127–138.

12 Siegel, M.J. and Evens, R.G. (1999). Advances in the use of computed tomography. *JAMA* 281 (14): 1252–1254.

13 Fatahi, M. and Speck, O. (2015). Magnetic resonance imaging (MRI): A review of genetic damage investigations. *Mutation Research/Reviews in Mutation Research* 764: 51–63.

14 Gallamini, A., Zwarthoed, C., and Borra, A. (2014). Positron emission tomography (PET) in oncology. *Cancers* 6 (4): 1821–1889.

15 Cherry, S.R., Sorenson, J.A., and Phelps, M.E. (2013). *Physics in Nuclear Medicine*. Society of Nuclear Medicine and Molecular Imaging.

16 Gambhir, S.S. (2002). Molecular imaging of cancer with positron emission tomography. *Nature Reviews Cancer* 2 (9): 683–693.

17 Kapoor, V., McCook, B.M., and Torok, F.S. (2004). An introduction to PET-CT imaging. *Radiographics* 24 (2): 523–543.

18 Catana, C., Drzezga, A., Heiss, W.-D., and Rosen, B.R. (2012). PET/MRI for neurologic applications. *Journal of Nuclear Medicine* 53 (12): 1916–1925.

19 Shen, D., Wu, G., and Suk, H.-I. (2017). Deep learning in medical image analysis. *Annual Review of Biomedical Engineering* 19 (1): 221–248.

20 Ghafoorian, M., Mehrtash, A., Kapur, T. et al. (2017). Transfer learning for domain adaptation in MRI: application in brain lesion segmentation. In: *Medical Image Computing and Computer Assisted Intervention-MICCAI 2017: 20th International Conference, Quebec City, QC, Canada, September 11–13, 2017, Proceedings, Part III 20*, 516–524. Springer.

21 Sinha, I.K., Verma, S., and Singh, K.P. (2024). FAM: adaptive federated meta-learning for MRI data. *Pattern Recognition Letters* 186: 205–212.

22 Esteva, A., Chou, K., Yeung, S. et al. (2021). Deep learning-enabled medical computer vision. *npj Digital Medicine* 4 (1): 5.

23 Razzak, M.I., Naz, S., and Zaib, A. (2018). Deep learning for medical image processing: overview, challenges and the future. In: *Classification in BioApps: Automation of Decision Making* (ed. S. Borra, N. Dey, and A.S. Ashour), 323–350. Springer.

24 Lamba, K. and Rani, S. Deep learning based autonomous model for detection and classification of brain tumor disease. *2023 Second International Conference on Informatics (ICI)*, pp. 1–6, IEEE, 2023.

25 Lamba, K. and Rani, S. (2023). Explainable artificial intelligence for deep learning models in diagnosing brain tumor disorder. In: *International Conference on Micro-Electronics and Telecommunication Engineering*, 149–159. Springer.

26 Kaissis, G.A., Makowski, M.R., Rückert, D., and Braren, R.F. (2020). Secure, privacy-preserving and federated machine learning in medical imaging. *Nature Machine Intelligence* 2 (6): 305–311.

27 Sheller, M.J., Edwards, B., Reina, G.A. et al. (2020). Federated learning in medicine: facilitating multi-institutional collaborations without sharing patient data. *Scientific Reports* 10 (1): 12598.

28 Larrazabal, A.J., Nieto, N., Peterson, V. et al. (2020). Gender imbalance in medical imaging datasets produces biased classifiers for computer-aided diagnosis. *Proceedings of the National Academy of Sciences* 117 (23): 12592–12594.

29 Johnson, J.M. and Khoshgoftaar, T.M. (2019). Survey on deep learning with class imbalance. *Journal of Big Data* 6 (1): 1–54.

30 Baumgartner, C. F., Koch, L. M., Tezcan, K. C., et al. (2018). Visual feature attribution using Wasserstein GANs. *Proceedings of the IEEE Conference on Computer Vision and Pattern Recognition*, pp. 8309–8319.

31 Mehrabi, N., Morstatter, F., Saxena, N. et al. (2021). A survey on bias and fairness in machine learning. *ACM Computing Surveys (CSUR)* 54 (6): 1–35.

32 Wang, J. (2023). The power of AI-assisted diagnosis. *EAI Endorsed Transactions on e-Learning* 8 (4): e3.

33 El-Shafai, W., Mahmoud, A.A., Ali, A.M. et al. (2024). Efficient classification of different medical image multimodalities based on simple CNN architecture and augmentation algorithms. *Journal of Optics* 53 (2): 775–787.

34 Finlayson, S. G., Chung, H. W., Kohane, I. S., and Beam, A. L. (2018). Adversarial attacks against medical deep learning systems. arXiv preprint arXiv:1804.05296.

35 Paschali, M., Conjeti, S., Navarro, F., and Navab, N. (2018). Generalizability vs. robustness: adversarial examples for medical imaging. arXiv preprint arXiv:1804.00504.

36 Ali, S., Li, J., Pei, Y. et al. (2022). A comprehensive survey on brain tumor diagnosis using deep learning and emerging hybrid techniques with multi-modal MR image. *Archives of Computational Methods in Engineering* 29 (7): 4871–4896.

37 Wang, J. (2023). Review of graph neural networks for medical image. *EAI Endorsed Transactions on e-Learning* 9: 4358.

38 Topol, E. (2019). *Deep Medicine: How Artificial Intelligence can Make Healthcare Human Again*. Basic Books, Hachette UK.

39 Van der Velden, B.H., Kuijf, H.J., Gilhuijs, K.G., and Viergever, M.A. (2022). Explainable artificial intelligence (Xai) in deep learning-based medical image analysis. *Medical Image Analysis* 79: 102470.

40 Lahat, D., Adali, T., and Jutten, C. (2015). Multimodal data fusion: an overview of methods, challenges, and prospects. *Proceedings of the IEEE* 103 (9): 1449–1477.

41 Wald, L.L., McDaniel, P.C., Witzel, T. et al. (2020). Low-cost and portable MRI. *Journal of Magnetic Resonance Imaging* 52 (3): 686–696.

42 Davids, J., Lidströmer, N., and Ashrafian, H. (2022). Artificial intelligence in medicine using quantum computing in the future of healthcare. In: *Artificial Intelligence in Medicine*, 423–446. Cham: Springer.

43 Kim, M.M., Parolia, A., Dunphy, M.P., and Venneti, S. (2016). Non-invasive metabolic imaging of brain tumours in the era of precision medicine. *Nature Reviews Clinical Oncology* 13 (12): 725–739.

8

Applications and Clinical Impacts of Biomedical Imaging

Divya Gupta, Jaspreet Kaur, and Sheenam Middha

Department of Computer Science and Engineering, Chandigarh University, Mohali, Punjab, India

8.1 Introduction

Biomedical imaging is one of the most important and fundamental aspects of modern healthcare. It gives academics and medical professionals a potent way to examine the human body without causing any harm. Through this technicalability,they can visualize interior structures, examine physiological and pathological processes, and get a thorough understanding of the anatomy and function of the body. Biomedical imaging is a multidisciplinary field that combines principles of biology, physics, engineering, and computer science [1]. Biomedical imaging enables a variety of clinical and research applications by leveraging cutting-edge technologies and a variety of imaging modalities, including X-rays, computed tomography (CT) scans, magnetic resonance imaging (MRIs), ultrasounds, and positron emission tomography (PET) scans. These tools are essential for precisely identifying illnesses, creating efficient treatment plans, and tracking the results of therapies over time. The execution of healthcare is greatly improved by the ability to acquire accurate and detailed scans of the body of an individual, which also improves patient outcomes. In order to make sure that medical decisions are accurate and personalized to each patient's needs, biomedical imaging is essential for detecting problems early on, directing medical treatments, and assessing the response to therapy [2, 3].

The diagnosis and treatment of a number of brain illnesses depend heavily on medical imaging methods, especially CT and MRI. High-resolution images from these cutting-edge technologies assist medical experts in identifying both structural and functional disorders in the brain, allowing for precise and prompt intervention. Hemorrhagic stroke, ischemic stroke, trauma, brain tumors, and skull fractures are among the prevalent brain disorders that can be identified by

Biomedical Imaging Technology: Signal Processing Strategies and Innovations, First Edition.
Edited by Ayush Dogra, Shalli Rani, and Ankita Sharma.

CT and MRI. When a blood artery in the brain bursts, blood seeps into the brain's surrounding tissue, resulting in a hemorrhagic stroke. On the other hand, an ischemic stroke happens when a blood clot blocks a blood vessel that supplies blood to the brain. Trauma is the term for significant physical harm or disruptions to the body, frequently brought on by violent crimes or accidents, which can also have psychological or emotional repercussions [4]. Both malignant and benign (non-cancerous) brain tumors are defined by the buildup of aberrant cells in the brain. These imaging methods help with precise diagnosis and efficient treatment planning by offering comprehensive insights into such illnesses. In addition to its current clinical and diagnostic capabilities, biomedical imaging continues to evolve rapidly with the integration of advanced computational techniques such as artificial intelligence (AI), machine learning (ML), and image processing algorithms. These technologies are enhancing image quality, automating detection of abnormalities, and enabling predictive analysis for earlier diagnosis. Furthermore, the development of hybrid imaging techniques—such as PET–CT and PET–MRI—provides a more comprehensive view by combining functional and structural data into a single scan. This fusion allows for a more accurate assessment of disease progression and response to treatment. The future of biomedical imaging is also geared toward greater personalization, where imaging data will be integrated with genetic, molecular, and clinical information to guide individualized treatment plans. As these innovations continue to mature, biomedical imaging will remain a cornerstone of precision medicine, significantly transforming the landscape of modern healthcare.

8.2 Essential Techniques in Biomedical Imaging

Biomedical imaging is founded on a core set of principles that enable the visualization of internal organs and physiological processes within the human body. These principles ensure the generation of accurate, detailed, and meaningful images for both clinical and research purposes. The foundational concepts of biomedical imaging encompass technological innovation, patient interaction, and safety considerations. Together, these elements facilitate the creation of high-quality, relevant images that support clinical diagnosis, medical research, and ongoing innovation [5, 6]. A variety of imaging technologies are employed in this field, each offering distinct benefits and applications. A comparative overview of these technologies is illustrated in Figure 8.1.

8.2.1 Computed Tomography (CT)

This advanced imaging technique uses specialized X-ray equipment to produce a number of incredibly detailed cross-sectional images, or scans, of inside body

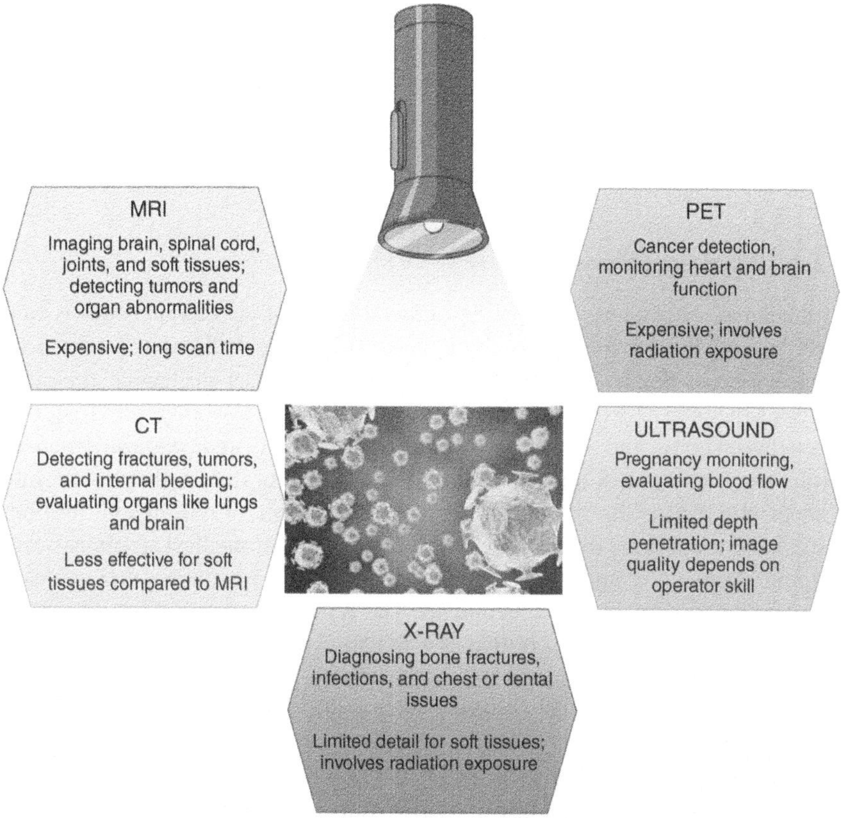

Figure 8.1 Comparison of various technologies in biomedical imaging domain.

components. This procedure is also known as computerized tomography or computerized axial tomography (CAT) scanning. CT scans provide a comprehensive view of various body regions, including blood vessels, soft tissues, and bones, by capturing images from various angles and transforming them into incredibly detailed cross-sectional slices. CT must be used in order to diagnose and assess cancer and other malignancies. It is commonly used to identify abnormal growths, assist in locating and characterizing tumors, and provide crucial information regarding the size, location, and stage of the cancer. CT scans assist in directing biopsy procedures by accurately identifying tumors for tissue harvesting [7]. Additionally, CT scans are used to monitor the progression of the illness, evaluate the effectiveness of treatment, and spot any recurrences. It is a helpful tool for both diagnosing and treating cancer. Despite the fact that CT scanning is a very helpful procedure in modern medicine, there are certain risks involved. Like other imaging techniques that employ radiation, like X-rays, CT scans can increase the risk of health issues caused by radiation. This includes the potential for cancer brought on by extended or severe radiation exposure. As a result, CT imaging is used with caution and the possible dangers are taken into account, particularly when high dosages or frequent use are involved. However, as long as appropriate safety precautions are taken, the advantages of CT imaging in terms of precise diagnosis, planning of therapy, andobservation of patients usually exceed the hazards.

8.2.2 Ultrasound Imaging

In medicine, ultrasound imaging is a crucial diagnostic technique that is frequently used to evaluate and examine a range of patient diseases. In essence, ultrasound uses high-frequency signals noises, which act as mechanical pressure waves as they travel through tissues. Echoes are sound waves that have frequencies between 20 Hz and 20 kHz. Depending on the density and composition ofthe tissue, the sound waves are reflected to differing degrees when they come into contact with it. The ultrasound gadget records the echoes, or reflected ultrasound waves, and transforms them into a visual depiction that medical specialists can examine. Depending upon the information required, many kinds of images can be created with sonographic instruments. The B-mode ultrasound image, which offers a two-dimensional descriptive view of tissue, is among the most often used types of ultrasound images. Numerous different specialized images can be produced by ultrasound imaging. For example, it may quantify tissue migration over time and display blood flow, giving important details about the body's dynamic activities [8]. Additionally, it can be used to identify blood arteries, identify certain chemicals, evaluate the flexibility of tissues, and even see the three-dimensional region's anatomy. Ultrasound is particularly helpful for real-time imaging, such as tracking blood flow or organ motion, and it can be used to precisely guide surgeries like biopsies or syringe placements.

8.2.3 Magnetic Resonance Imaging (MRI)

Remotely monitored magnetic resonance (MR) impulses are used in MRI, a sophisticated tomographic imaging technique, to produce accurate photographs of an object's underlying physical and chemical properties. MRI offers exceptional flexibility and precision in identifying the spatial characteristics of body tissues and structures, in contrast to many other imaging modalities. It works by building a diverse data collection that shows the distribution of specific physical attributes, such cell density or proton behavior, over different spatial locations. MRI is unique in that it can produce multiple sectional images at nearly any orientation, allowing for precise cross-sectional depictions of the human body. These segmented photos provide a detailed look inside and may be taken from any angle or depth. Three-dimensional volumetric pictures [9], which provide an expanded view of an area or organ by preserving its whole spatial volume, can also be produced by MRI. Complex anatomical structures can be more clearly seen and evaluated thanks to these multidimensional pictures. MRI allows physicians to gather important diagnostic information with excellent resolution and accuracy, whether it is in the form of images in two dimensions, three-dimensional in form volumes, or four-dimensional interpretations that incorporate time.

8.2.4 Positron Emission Tomography (PET)

This is a potent imaging technique that can be used to visualize the heart, brain, and cancer as well as for functional and molecular imaging. A single tracer is commonly used in traditional PET scans to offer precise functional or molecular information that is typically associated with disease. This implies that imaging various biological processes necessitates the use of numerous scans. For example, evaluating glucose metabolism using an 18F-fluorodeoxyglucose ([18F]FDG) PET scan and examining protein synthesis and identifying the boundaries of a brain tumor with a separate 11C-methionine ([11C]MET) scan. On the otherhand, multiplexed PET imaging allows for the simultaneous monitoring of several targets of interest from various tracers in a single scan. This method provides further data that can help clinicians make better decisions. Furthermore, it is beneficial to use many tracers in just one PET scan in order to shorten the examination duration and increase diagnostic efficiency. In multiplexed PET, different tracers are injected either simultaneously or consecutively with a few minutes' lag. Each tracer's dynamic or static data from imaging is then recovered in a single scan. The technique's proof-of-principle has been shown in preliminary research using cardiac imaging, simulated brain imaging, and phantom studies. However, the emission of photon pairs during positron absorption is a major obstacle to simultaneous multichannel PET imaging. Since these photons don't have distinct energy information, it's challenging to identify each photon pair's source. Consequently, the merged signals generated by the trace substances are detected by the PET scanner, making it harder to differentiate between them.

8.3 Applications of Biomedical Imaging

Biomedical imaging has a wide array of programs in healthcare, research, and medical exercise. It performs an important role in diagnosing diseases, tracking patient conditions, making plans remedies, and advancing scientific studies [10, 11]. Various applications in biomedical imaging domain have been presented in Figure 8.2.

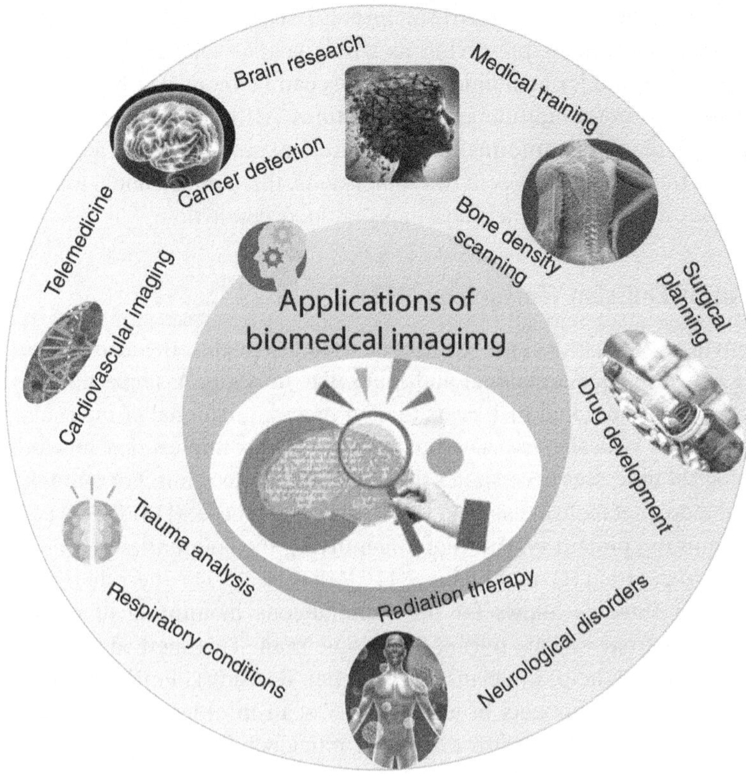

Figure 8.2 Applications in biomedical imaging domain.

8.3.1 Diagnostic Imaging

Diagnostic imaging plays a crucial role inside the early detection and diagnosis of diverse diseases. Biomedical imaging techniques, along with CT scans, MRI, and ultrasound, allow healthcare specialists to visualize inner structures and become aware of abnormalities.

8.3.1.1 Detection of Disease

Biomedical imaging is essential for detecting diseases at an early stage, when they are more treatable. Some of the diseases that can be detected using diagnostic imaging include:

- **Cancers**: Imaging techniques like CT scans, MRI, and ultrasound can assist pick out tumors and locate most cancers at an early level.

- **Cardiovascular Conditions**: Diagnostic imaging can assist discover cardiovascular diseases, such as coronary artery ailment, coronary heart failure, and stroke.
- **Neurological Disorders**: Imaging techniques like MRI and CT scans can assist diagnose neurological situations, which include multiple sclerosis, Parkinson's disorder, and mind tumors.
- **Musculoskeletal Problems**: Diagnostic imaging can help detect musculoskeletal issues, which include fractures, osteoporosis, and joint illnesses.

8.3.1.2 Radiology and Pathology

Radiologists and pathologists depend heavily on diagnostic imaging to examine internal systems and interpret the character of diseases. As an instance:

- **CT Scans for Brain Hemorrhages**: CT scans can assist detect brain hemorrhages, which may be lifestyle-threatening if left untreated.
- **MRI for Multiple Sclerosis**: MRI can help diagnose more than one sclerosis, a chronic and regularly disabling disorder.
- **Ultrasound for Joint Diseases**: Ultrasound can help diagnose joint sicknesses, including osteoarthritis and rheumatoid arthritis.

8.3.2 Treatment Planning

Treatment planning is an important step in the management of various medical conditions, and imaging plays a vital role in this process. Imaging modalities such as MRI, CT scans, and PET/CT provide us a valuable information that helps healthcare professionals to plan and execute effective treatment strategies [12].

8.3.2.1 Surgical Planning

Imaging plays a critical role in planning surgeries, especially for complex procedures such as brain, heart, or orthopedic surgeries. MRI and CT scans provide detailed images of tissues, organs, and blood vessels, helping surgeons plan more precisely way. Surgical planning involves:

- **Anatomical Mapping**: Imaging helps surgeons to create a detailed map of the patient's anatomy, including tissues, organs, and blood vessels.
- **Tumor Localization**: In cancer surgery, imaging helps identify the exact location and extent of tumor, enabling the surgeons to plan accurate resections.
- **Risk Evaluation**: Imaging helps the surgeons calculate the potential risks and complications, such as bleeding or damage to surrounding tissues.

8.3.2.2 Radiation Therapy

In cancer treatment, imaging needs radiotherapy for pin-pointing the exact location of tumors. Techniques such as PET/CT are often used to identify the best angles and dosages for delivering radiation to the malignant cells while minimizing damage to healthy tissues. Radiation therapy involves:

- **Tumor Localization**: Imaging helps in identifying the exact location and extent of tumor, enabling the oncologists to plan targeted radiation therapy.
- **Dose Planning**: Imaging helps the oncologists to determine the optimal radiation dose and delivery schedule.
- **Treatment Monitoring**: Imaging helps the oncologists in monitoring the effectiveness of radiation therapy and to adjust the treatment plans as needed.

8.3.3 Monitoring and Evaluation

Biomedical imaging plays an important role in monitoring and evaluating diseases, which enables the healthcare professionals to track the progression of disease and assess the effectiveness of treatment so that they can make informed decisions about patient care.

8.3.3.1 Disease Monitoring

Biomedical imaging helps to monitor the progression of disease over time, track the effectiveness of treatments, and helps to detect any changes in the condition of the patient. It is used to track the tumor shrinkage during chemotherapy or the response to cardiac treatments [13]. This enables healthcare professionals to:

- **Track Treatment Response**: Biomedical imaging helps to assess that how well treatments are working which enables the healthcare professionals to adjust the treatment plans accordingly.
- **Detect Changes in Condition**: Imaging helps to detect any changes in the condition of the patient which enables the healthcare professionals to respond promptly.
- **Monitor Disease Progression**: Biomedical imaging enables the healthcare professionals to track the progression of diseases, such as cancer, cardiovascular diseases, and neurological disorders.

8.3.3.2 Chronic Disease Management

In chronic conditions like arthritis, heart diseases, and neurological disorders, imaging is used to assess the condition regularly and to adjust the treatment plans accordingly. Role of imaging in chronic disease management is:

- **Regular Assessment**: Imaging is used to assess the condition regularly which enables the healthcare professionals to adjust treatment plans accordingly.

- **Monitoring Disease Progression**: Biomedical imaging enables healthcare professionals to track the progression of chronic diseases, enables prompt intervention.
- **Personalized Medicine**: Imaging enables the healthcare professionals to tailor the treatment plans to individual patients, taking into account their unique anatomy and physiology.

8.4 Clinical Impacts of Biomedical Imaging

Biomedical imaging has transformed modern healthcare by providing unprecedented insight into how the human body works. These imaging techniques offer a non-invasive, painless alternative to traditional surgery and have revolutionized the diagnosis, monitoring, and treatment of diseases. From advanced techniques such as MRI and positron emission tomography (PET) to more traditional X-rays and CT scans [14], biomedical imaging has opened the door to a wealth of possibilities in clinical care. The primary benefit of biomedical imaging is its ability to provide high-resolution, real-time images of organs, tissues, and other internal structures. These visual insights enable medical professionals to detect abnormalities such as tumors, blood clots, and organ dysfunction at the earliest possible stage, which is crucial for successful treatment. This technological advancement has improved treatment outcomes, allowing many diseases to be detected and treated before they cause irreversible damage.

8.4.1 Early Diagnosis

Biomedical imaging plays a key role in the early detection of numerous diseases, from cancer to cardiovascular disease [15, 16]. For diseases such as cancer, early diagnosis is especially important because tumors are often small and easily treatable in the early stages. Without imaging, these tumors can go undetected until they grow larger and are difficult to treat. This is where imaging techniques such as mammograms, CT scans, and MRIs become essential to detect the presence of the disease before symptoms appear. For example, in the case of breast cancer, mammograms are an invaluable screening tool that can detect small tumors even in the early stages, before patients or doctors feel a lump. Early detection allows for less invasive treatments such as local excision of the tumor, rather than more aggressive treatments such as chemotherapy. Additionally, modern imaging techniques such as 3D mammography can now detect tumors at an earlier stagewith greater accuracy. CT scans and MRIs have proven invaluable in the early detection of lung cancer, as they often spot small tumors or suspicious nodules long before patients experience symptoms. This proactive approach gives doctors the opportunity to intervene earlier, improving chances of survival and providing more treatment options. Cardiovascular diseases such as coronary artery disease can also benefit from earlier detection through imaging techniques. For example, a coronary CT angiogram can detect plaque buildup in the arteries that could lead to a heart attack. Early identification of high-risk patients may allow timely intervention, such as lifestyle changes, medications, or even stent placement, to reduce the risk of serious cardiovascular events. In neurological disorders such as Alzheimer's disease (AD), early detection with MRI can reveal brain changes that may indicate the onset of dementia, even before noticeable cognitive decline. Early identification of these changes allows for the administration of medications that can slow the progression of the disease, as well as lifestyle changes to maintain brain health.

8.4.2 Improved Treatment Planning

Imaging technologies provide clinicians with a wealth of information that can be used to tailor treatment plans to the individual needs of the patient. By allowing doctors to visualize the exact location, size, and nature of a disease or tumor, imaging ensures that treatment plans are accurate and effective. In cancer treatment, for example, imaging allows doctors to determine the best course of action by clearly defining the size and spread of tumors. Positron emission tomography (PET) and CT help doctors assess the metabolic activity of cancer cells [17], providing a clearer picture of how the cancer is progressing. This detailed information guides decisions about surgery, chemotherapy, or radiation therapy. By accurately determining the extent of the cancer, doctors can better plan how much tumor needs to be surgically removed and whether additional treatments are needed. The ability to locate tumors with imaging is also valuable in radiation therapy. Using 3D imaging, clinicians can accurately map the location tumor, ensuring that the radiation is directed precisely at the cancer cells while minimizing exposure to the surrounding healthy tissue. This approach not only improves the effectiveness of the treatment, but also reduces the side effects associated with radiotherapy, such as skin irritation and fatigue. In the field of neurosurgery, imaging technologies such as MRI and CT scans play a vital role in surgical planning for conditions such as brain tumors or spinal cord injuries. Detailed images allow surgeons to precisely navigate delicate areas of the brain or spine, reducing the risk of complications and improving patient outcomes. In fact, advances in intra operative imaging, which provide real-time imaging during surgery, have dramatically improved the accuracy of these procedures. Beyond oncology and neurosurgery, imaging plays a role in personalized treatment planning in many other medical specialties. In cardiology, for example, cardiac MRI can evaluate heart function, helping doctors determine whether a patient needs a pacemaker, valve replacement, or other procedures.

Similarly, in orthopedics, MRIs and X-rays help evaluate fractures orjoint injuries, guiding decisions about whether surgery or conservative treatment is best.

8.4.3 Monitoring and Assessment

The role of biomedical imaging extends beyond diagnosis and treatment planning, including monitoring and evaluation throughout the course of a disease. Whether monitoring the progress of cancer treatment or assessing a patient's recovery after surgery, imaging provides clinicians with real-time feedback that is essential to adjust treatment plans and ensure the best possible results.In cancer, images are used not only for diagnosis, but also to monitor the patient's response to the treatment. For example, a patient undergoing chemotherapy or radiation therapy may have periodic examinations to monitor the tumor's response to treatment. If the tumor shrinks, the treatment plan can continue as planned, but if it does not respond or grows, the doctor may consider changing to another form of treatment.MRI and CT scans are also essential for evaluating the effectiveness of treatments for neurological diseases such as stroke or multiple sclerosis. In stroke patients, imaging is used to determine the extent of brain damage and to monitor recovery over time. Similarly, in patients with multiple sclerosis, MRI can track the progression of brain and spinal cord damage, guiding treatment adjustments to slow the progression of the disease.The use of imaging for monitoring extends to chronic diseases such as arthritis and diabetes. MRIs and X-rays are commonly used to assess damage caused by arthritis, including joint degeneration and inflammation. Regular imaging can help doctors make more informed decisions about pain management, physical therapy, or surgical procedures, improving a patient's quality of life.Additionally, imaging is an essential part of postoperative care. After surgery, especially in high-risk procedures such as joint replacements or organ transplants, imaging is used to assess the healing process. X-rays, CT scans, and ultrasounds can detect complications such as infections, deformities, or fluid buildup, allowing doctors to take action before the situation becomes more serious.

8.4.4 Minimizing Invasive Procedures

One of the foremost critical commitments of biomedical imaging is the reduction of obtrusive strategies. Within the past numerous therapeutic analyze and medicines required surgeries or biopsies which carried dangers of disease dying and long recuperation times. Be that as it may with progressed imaging advances these methods have gotten to be distant less common as imaging allows clinicians to accumulate crucial data noninvasively. Within the domain of diagnostics imaging advances like ultrasound CT and MRI filters empower specialists to recognize inner anomalies such as tumors breaks or diseases without having to perform intrusive exploratory surgeries.This minimizes persistent hazard and inconvenience making healthcare more patientfriendly and less traumatic. For case in cases of suspected cancer a biopsy utilized to require a specialist to evacuate a tissue test from the tumor which included entry points and anesthesia. Todayimaging-guided biopsy methods utilize ultrasound or CT checks to pinpoint the precise area of the irregular tissue permitting specialists to extricate a tissue test with a fine needle. This decreases the requirement for obtrusive surgeries and permits for a speedier recuperation all whereas keeping up demonstrative precision In expansion to demonstrative strategies imaging plays a part in guiding helpful intercessions. In cardiology for occurrence catheter-based methods like angioplasty can be performed with the assistance of real-time X-ray direction which makes a difference specialists embed stents or clear blockages without the requirement for openheart surgery. In orthopedics negligibly intrusive surgeries guided by imaging innovations such as MRI permit specialists to repair bone breaks or torn tendons with little entry points driving to less torment and quicker recuperating times [18].

8.4.5 Accelerating Research and Innovation

Biomedical imaging is important not just for treating patients, but also for improving medical research and development. It helps scientists see how diseases happen in realtime and with great accuracy. For example, in cancer research, techniques like PET and MRI are used to study how tumors grow and spread. These tools allow researchers to observe how tumors behave and respond to different treatments. This information is essential for creating targeted therapies that are more effective and have fewer side effects than traditional treatments like chemotherapy. Moreover, imaging methods like fMRI and PET have changed how we understand brain disorders. Researchers can now watch brain activity in realtime and see which parts are activated during various tasks or in response to medications. This has led to new treatments for diseases like Alzheimer's, Parkinson's, and schizophrenia. In cardiology, imaging techniques are used to examine heart function and blood flow, which helps develop better treatments for heart disease. Imaging has also contributed to advancements in medical devices like stents and pacemakers, used to treat conditions like coronary artery disease and irregular heartbeats. Another exciting development is the use of AI with biomedical imaging. AI is increasingly used to analyze large amounts of medical images, helping to find patterns, predict how diseases progress, and assist in early detection of illnesses. This could greatly improve diagnostics, making them faster and more accurate.

8.4.6 Cost Efficiency

Although biomedical imaging technologies are often expensive, they have proven to be very costeffective in the long run. By enabling early diagnosis and reducing the need for invasive procedures, imaging technologies can help reduce overall healthcare costs. Early detection of diseases such as cancer, cardiovascular disease, and neurological disorders allows for more targeted and less invasive treatments, which are generally less expensive than the more aggressive interventions required in advanced stages of the disease. For example, treating early-stage cancer with surgery or radiotherapy is usually less expensive than chemotherapy,which is required for advanced cancer. Imaging also helps reduce the need for multiple diagnostic tests, which can be timeconsuming and expensive. Instead of performing a series of invasive procedures, doctors can use imaging to gather all the necessary information in a single exam. This not only saves money, but also reduces the burden on the healthcare system.In addition, imaging allows for better allocation of resources. By providing clear information about a patient's condition, clinicians can avoid unnecessary treatments or hospitalizations. This optimizes the use of healthcare resources, thereby reducing costs for patients and providers [19].

8.5 Case Studies

Early diagnosis, planning for therapy, and illness tracking are made possible by biomedical imaging, which is essential to clinical practice. The case studies that follow demonstrate the significance of several imaging modalities in practical clinical uses:

8.5.1 Advanced PET/CT Imaging for Tracking Cancer Metastases

Even though the patient had previously had first chemotherapy, more imaging was required to assess the severity of the condition and the treatment's effectiveness. Two different tracers were employed in the PET/CT scans: 18F-FDG and 11C-MET. FDG is commonly used in PET imaging to assess glucose metabolism, which offers insight into the tumor's metabolism, whereas MET is used to scan protein synthesis, which offers data on the tumor's propagation and survivability.

In this sophisticated case study, a 58-year-old woman with non-small-cell lung cancer (NSCLC) had her cancer metastases monitored using PET/CT hybrid imaging. Although the sufferer had already had first chemotherapy, further imaging was necessary to determine the severity of the illness and the efficacy of the treatment. The PET/CT scans used two distinct tracers: 11C-MET and 18F-FDG. While MET is employed to image the synthesis of proteins, which provides information on the tumor's proliferation and survivability, FDG is frequently utilized in PET imaging to evaluate glucose metabolism, which provides insight into the tumor's metabolism. The initial FDG-PET/CT scan revealed a primary tumor in the right lung and possible lymph node involvement, leading to the possibility that the disease has disseminated. However, a follow-up scan with MET tracer provided more crucial information. The scan demonstrated a significant rise in the tumor's protein production activity and identified areas of liver metastases that had not been seen in the initial scan. With this knowledge, clinicians were better equipped to understand the full extent of the condition and adjust the treatment strategy. Based on the extensive functional and anatomical data provided by the combined PET/CT imaging, a personalized treatment plan was developed for the patient. As portion of the treatment, targeted medications were employedto reduce the tumor's metabolic activity. Follow-up PET/CT scans were used to monitor the patient's response to the treatment. The scans showed a significant drop in metabolic activity, indicating that the tumor was responding to the therapy. Furthermore, the scans confirmed cancer shrinkage, proving the treatment's efficacy.

The use of advanced hybrid PET/CT imaging proved to be quite helpful in this case. By allowing physicians to simultaneously observe the cancer's structural and functional characteristics, it allowed them to gain a comprehensive understanding of the tumor's behavior and spread. This dual capability allowed for the precise localization of metastases, which is essential for accurate staging and treatment planning. Additionally, timely adjustments to the therapeutic approach were made possible by continuous tracking of the tumor's response to treatment, which enhanced the patient's care. The use of different tracers, such FDG and MET, gave a more complete picture of the tumor's metabolism and proliferative characteristics, which significantly increased the diagnostic accuracy. Ultimately, this state-of-the-art imaging technique was crucial to improving the patient's prospects.

8.5.2 High-Resolution MRI for the Prompt Identification of Alzheimer's Disease

The individual, who was a 72-year-old male, had early signs of cognitive decline and a family history of AD. The level of neurotoxicity was assessed using sophisticated imaging methods, including high-resolution 3T MRI and amyloid PET imaging. On the 3T MRI, the hippocampus displayed mild cortical thinning, which is one of the earliest indications of neurodegeneration linked to AD. The accumulated PET scan, which showed a significant accumulation of amyloid plaque in the brain, a hallmark of AD, provided more evidence of the disease's existence. Long before the clinical signs of severe cognitive decline were fully apparent, the combination of data from structural MRI with operational amyloid imaging using Pet allowed for an early and precise diagnosis of AD.

Following confirmation of the diagnosis, the patient was given an antibiotic to help control symptoms and enhance cognitive performance. Additionally, lifestyle changes were recommended to delay the progression of the disease. Among these adjustments were an emphasis on physical exercise and cognitive therapy. These cutting-edge imaging methods made it possible for early discovery, which allowed for medical treatment at a point when the disease might be better controlled. A more individualized treatment approach was made possible by the combination of the structural MRI and aggregate PET imaging, which also assisted in diagnosis confirmation. In order to improve the person's standard of life, reduce the advancement of AD, and maybe postpone the emergence of more severe symptoms, early diagnosis and management are essential.

8.5.3 Breast Shape Analysis

In both cosmetic and reconstructive surgery, determining the contour of the breasts is an essential stage in assessing the outcome. Since there are currently no widely recognized methods for analyzing breast shape, surgeons still typically perform visual assessments. One major scientific problem is creating trustworthy techniques to accurately evaluate both natural and artificial breast forms. Although a number of reliable measurement methods have been put out in the literature, none of them have found widespread use in clinical settings. While some studies advise using manual linear measurements and anatomic reference points to analyze breast shape, others advise using techniques like the displacement of water, MRI, X-rays, or thermoplastic material molding to quantify volume. When examining patients, surgeons can recognize a variety of anatomical sites and features of interest by using their experience. Nevertheless, the instruments for gathering measurements—such as the length of paths across the breast surface, breast surface area, and comparable metrics—are frequently insufficient or nonexistent. Additionally, the subject's posture and other experimental parameters have a significant impact on manual measurements. In an effort to find a workable solution, a multidisciplinary team of physicians from the National Cancer Institute (INT) and computational scientists from the University of Catania recently investigated novel technical methods for breast form analysis. In order to enable plastic surgeons objectively assess the results of reconstructive surgery, this endeavor focuses on defining and analyzing breast shape using clinical semantics. In order to direct their study, the research team established a number of important goals, especially those that were pertinent to clinical practice:

- Breast component segmentation and characterization.
- Distances between landmark locations on the female thorax can be measured as lengths of path over the body's surface or as line distances in three dimensions.
- Areas, volumes, and the proportion of each unit to the total area are measured.
- Angles adjacent landmark planes are measured.

As area ratios between the breast's lower and upper poles, as well as other clinically relevant surfaces, are frequently used in reconstructive surgery, these measures are essential. Accurate assessment also depends on the separations between anatomical landmarks. Surgeons have identified the crucial problem of giving precise, unbiased meanings to ambiguous but frequently used terminology like "sweet slope" and "inferior pole." This has been addressed by the successful use of fake colors to visualize curvature and convexity. Key objectives in breast reconstruction surgery include symmetry and an ideal breast shape, which can be attained by surgeons through the analysis of curvature and convexity.

8.6 Conclusion

Since biomedical imaging offers precise, in-depth, and non-invasive insights into the organization and function of the human body, it has completely transformed both biomedical research and healthcare. Its applications cover a wide range of medical specialties and range from early disease identification to personalized therapy planning. Professionals are now able to precisely diagnose complex disorders, track the course of diseases, and assess the effectiveness of treatments thanks to advanced imaging modalities including MRI, CT, X-ray ultrasound, and PET. The range and precision of biomedical imaging are also being improved by new technologies including molecular imaging, AI-powered image processing, and hybrid imaging systems. By facilitating early intervention and efficient use of resources, these developments not only enhance clinical results but also lower healthcare expenses. Since biomedical imaging connects cutting-edge technology with patient-centered care, its clinical significance is indisputable. Biomedical imaging will surely become more and more important in tackling global health issues, influencing the direction of medicine, and enhancing patient quality of life everywhere as technology and research advance.

Biomedical imaging's future depends on combining state-of-the-art technology with multidisciplinary research to overcome present constraints and open up new avenues. Future studies in these fields will improve biomedical imaging's capacity for diagnosis and treatment while also advancing patient-centered, affordable, and easily accessible healthcare systems.

References

1 Weissleder, R. and Nahrendorf, M. (2015). Advancing biomedical imaging. *Proceedings of the National Academy of Sciences* 112 (47): 14424–14428.

2 Ben Yedder, H., Cardoen, B., and Hamarneh, G. (2021). Deep learning for biomedical image reconstruction: a survey. *Artificial Intelligence Review* 54 (1): 215–251.

3 Seo, H., Badiei Khuzani, M., Vasudevan, V. et al. (2020). Machine learning techniques for biomedical image segmentation: an overview of technical aspects and introduction to state-of-art applications. *Medical Physics* 47 (5): e148–e167.

4 Lin, H., Wang, C., Cui, L. et al. (2022). Brain-like initial-boosted hyperchaos and application in biomedical image encryption. *IEEE Transactions on Industrial Informatics* 18 (12): 8839–8850.

5 Nazir, S., Dickson, D.M., and Akram, M.U. (2023). Survey of explainable artificial intelligence techniques for biomedical imaging with deep neural networks. *Computers in Biology and Medicine* 156: 106668.

6 Rajeyyagari, S., Hung, B.T., and Chakrabarti, P. (2022, February). Applications of artificial intelligence in biomedical image processing. *2022 Second International Conference on Artificial Intelligence and Smart Energy (ICAIS)*, 53–57. IEEE.

7 Moul, J.W., Shore, N.D., Pienta, K.J. et al. (2024). Application of next-generation imaging in biochemically recurrent prostate cancer. *Prostate Cancer and Prostatic Diseases* 27 (2): 202–211.

8 Varoquaux, G. and Cheplygina, V. (2022). Machine learning for medical imaging: methodological failures and recommendations for the future. *NPJ Digital Medicine* 5 (1): 48.

9 Sun, Z., Ng, K.H., and Ramli, N. (2011). Biomedical imaging research: a fast-emerging area for interdisciplinary collaboration. *Biomedical Imaging and Intervention Journal* 7 (3): e21.

10 Chakraborty, S. and Mali, K. (2020). An overview of biomedical image analysis from the deep learning perspective. In: *Applications of advanced machine intelligence in computer vision and object recognition: emerging research and opportunities* (ed. S. Chakraborty and K. Mali), 197–218. New York: IGI Global.

11 Fernando, K.R.M. and Tsokos, C.P. (2023). Deep and statistical learning in biomedical imaging: state of the art in 3D MRI brain tumor segmentation. *Information Fusion* 92: 450–465.

12 Buaka, E.S.D. and Moid, M.Z.I. (2024). AI and medical imaging technology: evolution, impacts, and economic insights. *The Journal of Technology Transfer* 49 (6): 2260–2272.

13 Iqbal, A., Sharif, M., Yasmin, M. et al. (2022). Generative adversarial networks and its applications in the biomedical image segmentation: a comprehensive survey. *International Journal of Multimedia Information Retrieval* 11 (3): 333–368.

14 Ben Yedder, H., Cardoen, B., and Hamarneh, G. (2021). Deep learning for biomedical image reconstruction: a survey. *Artificial Intelligence Review* 54 (1): 215–251.

15 Reinke, A., Maier-Hein, L., and Müller, H. (2021). Common limitations of performance metrics in biomedical image analysis. *Proceedings of the Medical Imaging with Deep Learning (MIDL 2021).*

16 Zubair, M. (2023). Machine learning based biomedical image analysis and feature extraction methods. *Journal of Applied and Emerging Sciences* 13 (1): 31–39.

17 Zhao, T., Gu, Y., Yang, J. et al. (2024). BiomedParse: a biomedical foundation model for image parsing of everything everywhere all at once. arXiv preprint arXiv:2405.12971.

18 Reinke, A., Tizabi, M.D., Baumgartner, M. et al. (2024). Understanding metric-related pitfalls in image analysis validation. *Nature Methods* 21 (2): 182–194.

19 Levac, B., Jalal, A., and Tamir, J.I. (2023, April). Accelerated motion correction for MRI using score-based generative models. *2023 IEEE 20th International Symposium on Biomedical Imaging (ISBI)*, 1–5. IEEE.

9

Emerging Technologies and Innovations in Medical Imaging

Puneet Bawa[1] *and Manisha Rajput*[2]

[1] *School of Computer Science and Engineering, IILM University, Gurugram, India*
[2] *Chitkara University Institute of Engineering & Technology, Chitkara University, Punjab, India*

9.1 Introduction

Medical imaging refers to the use of technology to create visual representations of the internal body structures for clinical analysis and medical intervention. The field of medical imaging serves as a cornerstone of modern healthcare by enabling non-invasive diagnosis alongside the benefits of treatment planning and monitoring of diseases. The field of medical imaging utilizes different types of modalities such as X-rays [1], ultrasound [2], magnetic resonance imaging (MRI) [3], computed tomography (CT) [4], mammography [5], and nuclear imaging [6] with an effort of capturing the detailed anatomical and functional information corresponding to an individual. X-rays are a type of medical imaging that uses ionizing radiation to create images of bones and soft tissues which are used to diagnose conditions, mainly including bone fractures [7, 8], arthritis [9], and breast cancer [10]. The ultrasound uses high-frequency sound waves to create images of the inside of the body and is mainly used to diagnose conditions like heart and blood vessel abnormalities [11, 12] as well as to monitor a fetus during development [13]. Further, the most widely utilized modality of MRI utilizes radio waves and magnetic fields with an objective of creating detailed cross-sectional images of the body. MRIs are often utilized to create images of the brain [14], spinal cord [15] and blood vessels [16] such that this modality is observed to be painless and safe and can also be used for pregnant women and infants [17, 18]. CT uses a series of X-rays and a computer to create a comprehensive image of the body where the modality is widely utilized to diagnose conditions like cancer and to detect internal trauma. In like manner, mammography has been widely used by the researchers for the early identification of conditions like breast cancer [19, 20]. Additionally, the use

Biomedical Imaging Technology: Signal Processing Strategies and Innovations, First Edition.
Edited by Ayush Dogra, Shalli Rani, and Ankita Sharma.

of another kind of modality of nuclear imaging has been used in various research studies [21, 22]. Nuclear imaging is a method of producing images by detecting radiation from different parts of the body after administering a radioactive tracer to the patient. The images are then digitally generated on a computer and are interpreted by a nuclear medicine physician with an effort of making a diagnosis. Overall, Table 9.1 details the comparison of rising modalities using immersive technologies mainly including Extended Reality (XR)-Integrated Imaging alongside the state-of-the-art modalities discussing their advantages and limitations.

9.1.1 Background

The immersive technologies are advanced systems that blend the physical and digital worlds to create engaging and interactive experiences. These technologies encompass the three major key forms including:

- **Augmented Reality (AR)**: AR overlays digital information onto the real world which has enhanced real-world experiences with computer-generated visuals. This technology is commonly experienced through the usage of smartphones [25] or AR glasses [26]. This allows the users to interact with both digital and physical environments simultaneously.
- **Virtual Reality (VR)**: VR immerses users in a fully simulated 3D environment which is often achieved through headsets that replace the real world with a virtual one. VR is widely used in various fields including healthcare [27], education [28], and design [29] by providing users with a sense of presence in a non-physical space.
- **Mixed Reality (MR)**: This technology combines elements of both AR and VR which allows digital and physical objects to coexist and interact in real time [30, 31]. The use of this technology creates a more integrated experience combining both the benefits of AR and VR where the users have the capability of manipulating and engaging with both virtual and real-world elements.

These three key forms are collectively referred to as XR as a technology which is transforming various sectors of interest through the immersive experiences by enhancing understanding, engagement, and interaction of an individual to the real-word problem [32]. XR has played a pivotal role in revolutionizing the healthcare system by focusing on medical imaging which has enabled practitioners to visualize and interact with complex anatomical structures in three dimensions. The past decade has witnessed a significant surge in the technology which has led to the exploration of applications resulting in the increasing recognition of

Table 9.1 Comparative analysis of medical imaging modalities with key advantages and limitations.

Imaging modality	Non-invasive	High resolution	Real-time imaging	Functional imaging	Affordable	Radiation-free	Best suited for
X-ray [7–10]	✓	✓	×	×	✓	×	Bone fractures, chest imaging
CT scan [19, 20]	✓	✓✓	×	✓	×	×	Internal injuries, cancers
MRI [14–18]	✓	✓✓	×	✓	×	✓	Soft tissues, brain imaging
Ultrasound [11–13]	✓	✓	✓✓	×	✓✓	✓	Fetal imaging, organ evaluation
Mammography [19, 20]	✓	✓✓	×	×	✓	×	Breast cancer detection
Fluoroscopy [23]	✓	✓	✓✓	×	×	×	Dynamic studies, GI tract
Nuclear imaging [21, 22]	✓	✓	×	✓✓	×	×	Organ function, cancer diagnosis
XR-integrated imaging [24]	✓	✓✓	✓✓	✓	✓	✓	Training, real-time diagnostics

potential to improve patient outcomes and healthcare delivery [33]. In medical education, the immersive technologies have been utilized and studied by the researchers with an objective of simulating realistic clinical scenarios which has resulted in the availability of a risk-free environment for students to practice and hone their skills [34–36]. In like manner, surgeons are increasingly adopting immersive technologies utilizing the applications of XR for preoperative planning and intraoperative guidance [31, 37]. As a result, there is a need for an in-depth analysis that can yield valuable insights into the trends and patterns within this emerging body of research, along with guiding future research paths and facilitating the immersive implementation of these technologies in the field of medical imaging.

9.1.2 Contribution

This chapter provides a comprehensive bibliometric analysis of the publication landscape in medical imaging technologies specifically focusing on integration of immersive technologies with XR applications. The following list of contributions provides a thorough technical and practical discussion of the impact of publication trends and types on the advancement of medical imaging technologies:

- Identified temporal patterns in publication output highlighting periods of significant growth or decline.
- Provided quantitative metrics related to citations, h-index and total link strength with an effort of measuring the impact of authors, journals, and institutions.
- Highlighted key contributors driving research to enhance the recognition of significant collaborators and networks for potential partnerships.
- Mapped relationships between frequently used keywords to identify core research themes and thematic clusters.
- Analyzed collaboration patterns among the countries to foster interdisciplinary and international research efforts.
- Evaluated new and emerging areas of research by analyzing keyword density and recent publication trends.

9.1.3 Organization

The remainder of the chapter has been organized as follows: Section 9.2 details the methodology which involves selecting the papers based on the query with detailed inclusion and exclusion criteria. Further, the results of bibliometric analysis are detailed in Section 9.3 showcasing key indicators and trends regarding the emerging technologies and innovation in medical imaging. Section 9.4 discusses the discussions and limitations related to the emerging trends in medical imaging with the conclusion in Section 9.5.

9.2 Methodology

The bibliometric methodology concerning emerging technologies' role in medical imaging encapsulates the application of quantitative techniques on bibliometric data. Likewise, the SCOPUS database has been utilized to extract and analyze relevant scholarly works based on the search query as "*(TITLE-ABS-KEY ("augmented reality" OR "virtual reality" OR "extended reality" OR "mixed reality" OR "XR") AND TITLE-ABS-KEY ("medical imaging" OR "diagnostic imaging" OR "radiology"))*." The query leads to the technology evolution that has evolved rapidly in immersive reality technologies.

Figure 9.1 showcases the PRISMA flowchart used to visually represent the study selection process such that the query has resulted in a total number of publications (N = 5627). Further, there were publications (N = 81) with empty author

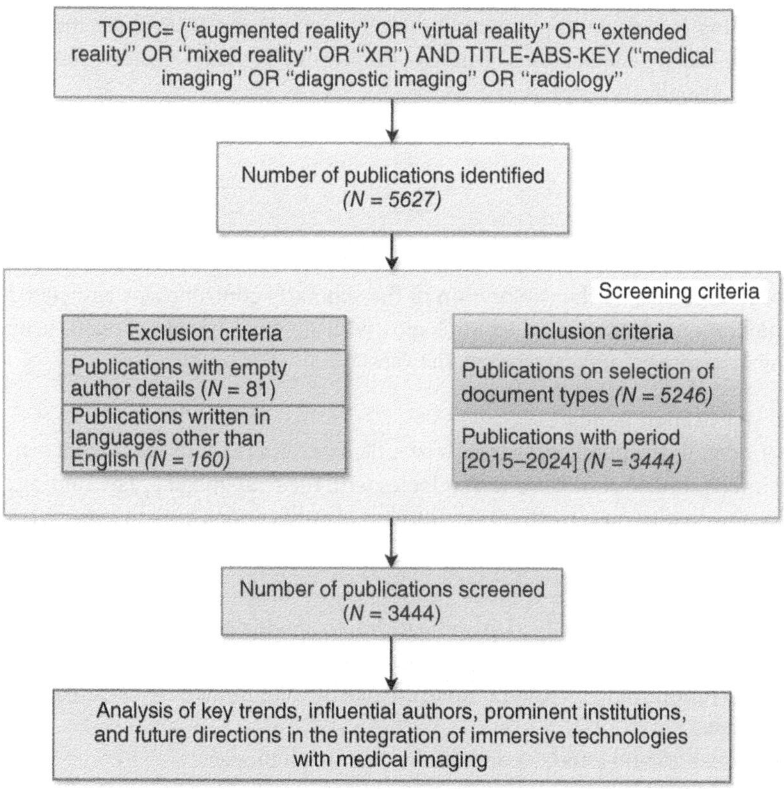

Figure 9.1 PRISMA flowchart for the selection methodology for the bibliometric analysis concerning the integration of immersive technologies with medical imaging.

details which were removed from the analysis due to the missing data. Likewise, there were (N = 160) publications that were in languages other than English. Also, the inclusion criteria for the study were based on the selection of document types (Article, Book, Book Chapter, Conference, Editorial, and Review) and the period (2015–2024) which resulted in finalized records (N = 3444) for conducting the bibliometric analysis. Overall, the research employs bibliometric analysis to drive the insights into key trends and opportunities concerning the integration of immersive technologies with medical imaging. The analysis has been conducted using VOSviewer and Biblioshiny [38] offering robust features for mapping and interpreting trends, collaborations and emerging topics in research. These tools offer dynamic and visually engaging representations that enhance the interpretation of bibliometric findings effectively. The VOSviewer has been utilized in the research such that the software supports large-scale datasets for comprehensive analysis. The software provides customizable parameters for node and link visualization. In like manner, Biblioshiny offers a comprehensive suite of metrics, including H-index and productivity trends. Overall, the VOSviewer focuses on network-based visualizations, whereas Biblioshiny provides detailed metrics and diverse visualization options.

9.3 Analysis

9.3.1 Research Dynamics

This section focuses on the exploration of the scholarly contributions associated with integration of immersive technologies with medical imaging research by focusing on temporal and categorical analyses.

9.3.1.1 Publication Trends

Figure 9.2 details the publication trends over the last decade (2015–2024) concerning the integration of immersive technologies with medical imaging. The findings illustrate the continuous growth in the number of publications from the year 2015 to 2020. However, a sharp rise in publications was observed in 2021 which is likely to be driven by the following reasons:

- Surge in the interest of medical imaging technologies during the COVID-19 pandemic.
- Increased funding and focus on remote diagnostics and immersive visualization technologies.
- Global collaboration efforts to address healthcare challenges.

Further, the dip can be observed in 2022 and 2023 which may be due to the saturation of AR/VR themes or may be forced by the slower adoption of newer XR innovations. The major possible challenge can be due to the reduced global research due to funding shifted toward recovery efforts post-pandemic. On the

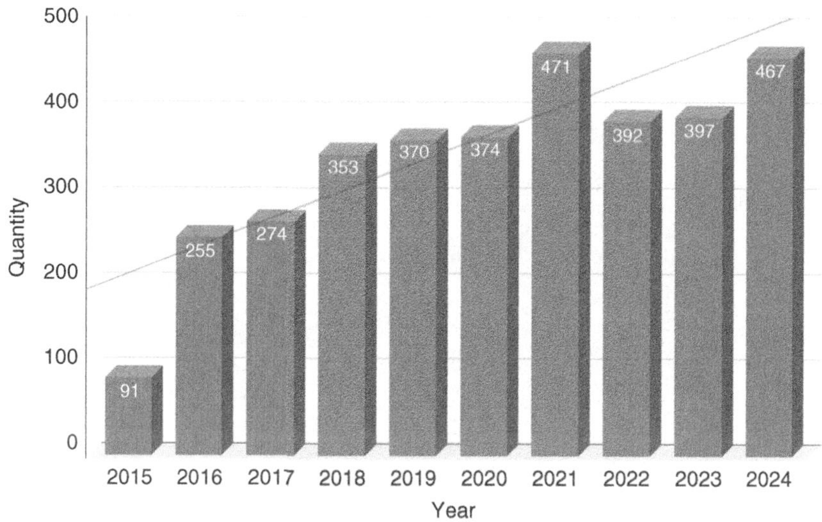

Figure 9.2 Publication trends over the last decade concerning the integration of immersive technologies with medical imaging.

other hand, there has been a strong rebound in 2024 with findings indicating the near surpassing of 2021. This has been due to the renewed interest in XR-driven innovations for personalized and AI-enhanced medical imaging. In recent times, there have been expanding applications of mixed reality which are being utilized by researchers in clinical practice and training. Additionally, the 2024 rebound suggests emerging opportunities for commercialization and clinical adoption of XR technologies. Overall, the trendline reflects the sharp rise indicating cyclical patterns of research interest over time in emerging technologies.

9.3.1.2 Publication Types

Figure 9.3 illustrates an analysis of diverse document types retrieved from the Scopus database over the last decade highlighting the integration of immersive technologies with medical imaging. The following are the details of the document types:

- **Article (71.1%)**: Peer-reviewed studies presenting original findings.
- **Conference Paper (18.8%)**: Contributions focused on emerging technologies and prototype developments.
- **Review (6.9%)**: Summaries of existing research and identification of future directions.
- **Editorial (1.5%)**: Insights and opinions on trends and challenges in the field.
- **Book Chapter (1.6%)**: In-depth explorations of specific topics.

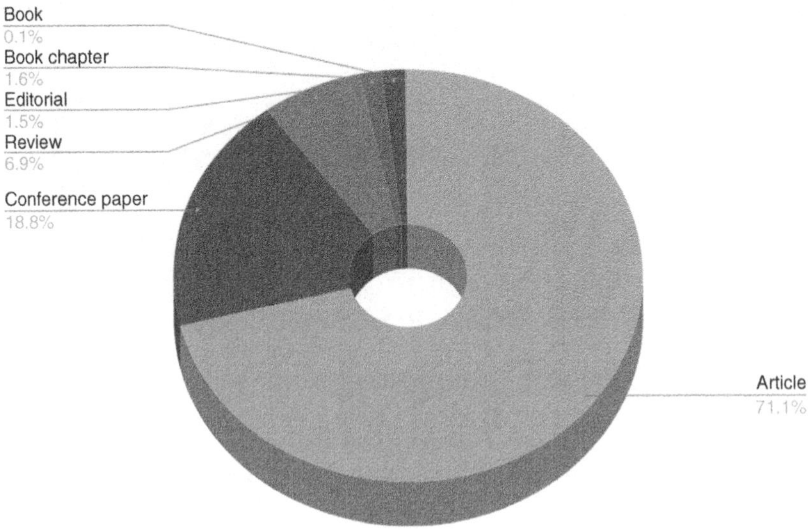

Figure 9.3 Analysis of publication types over the last decade concerning the integration of immersive technologies with medical imaging.

9.3.1.3 Country Impact Analysis

Table 9.2 details the country impact analysis focuses on identifying the contributions of different countries to the field of emerging immersive technologies in medical imaging. This analysis highlights the global distribution of research activity listing the total citations and average article citations for understanding the consistent innovation in the field. The United States ranks as the most prominent country in this domain with the highest number of citations with highest Average Article Citations (AAC = 30.8). China (AAC = 15.1) ranks second in the impact reflecting rapid growth in research and development but with nearly 1/3rd of total citations compared to the United States. The European countries including Italy (AAC = 18.5), France (AAC = 22.4), and Germany (AAC= 16) are emerging as global hubs with greater AAC in comparison to China highlighting innovation and large-scale studies. However, India's position at the bottom of the table might seem less prominent compared to global leaders like the USA and China. However, India has an equivalent (AAC = 15.1) compared to China highlighting the steadily advancing healthcare and technology domains alongside the increasing contributions to affordable and accessible solutions.

9.3.2 Key Contributors

This section highlights the key contributions by analyzing the impact of citations, authors, and journals in the field of emerging technologies in medical imaging.

Table 9.2 Overview of the country impact analysis based on the total citations concerning the emerging immersive technologies in medical imaging.

Country	Total citations	Average article citations
United States of America	19,013	30.8
China	7137	15.1
Italy	3612	18.5
France	3242	22.4
Germany	3145	16
United Kingdom	2169	16.4
Japan	1874	11.6
Canada	1495	16.8
Australia	1486	22.9
India	1226	15.1

The section aims at showcasing the most influential work that has shaped the domain with an effort to provide insights into future research directions.

9.3.2.1 Most Cited Papers

Table 9.3 helps in the identification of the most influential papers in the field of emerging technologies in medical imaging by examining their total citations per year (TC/year). Initially, Pettersen et al. [39] showcased the impactful works that led to the introduction of ChimeraX as the next-generation interactive visualization program. The research conducted has acted as the support for new areas mainly including light-sheet microscopy [40] and medical imaging data. Further, Minaee et al. [41] have conducted a comprehensive survey concerning image segmentation which has been found to play a prominent role in the field of AR. In like manner, the review has been conducted by Mo et al. [42] considering the state-of-the-art technologies of semantic segmentation based on deep learning with the research highlighting the field of AR. The research conducted by Jirsa et al. [43] has the potential to incorporate immersive technology for tailoring treatments for conditions like bitemporal epilepsy.

9.3.2.2 Author Impact Analysis

Table 9.4 details the analysis by highlighting the contributions of the top 10 authors in the field of emerging immersive technologies for medical imaging. The author NAVAB NASSIR has the greatest h-index (17) with the number of

Table 9.3 Overview of the most cited papers based on the total citations concerning the emerging immersive technologies in medical imaging.

Publication detail	Doi	TC	TC/year	Normalized TC
PETTERSEN EF, 2021, PROTEIN SCI	10.1002/pro.3943	4123	1030.75	183.62
MINAEE S, 2022, IEEE TRANS PATTERN ANAL MACH INTELL	10.1109/TPAMI.2021.3059968	1719	573	115.92
RIVENSON Y, 2019, NAT BIOMED ENG	10.1038/s41551-019-0362-y	422	70.33	17.79
STEVENS PE, 2024, KIDNEY INT	10.1016/j.kint.2023.10.018	395	395	222.78
MO Y, 2022, NEUROCOMPUTING	10.1016/j.neucom.2022.01.005	339	113	22.86
SHEPHERD JA, 2017, BONE	10.1016/j.bone.2017.06.010	305	38.13	9.19
SONKUSARE S, 2019, TRENDS COGN SCI	10.1016/j.tics.2019.05.004	286	47.67	12.06
AL-SHEIKH M, 2017, INVEST OPHTHALMOL VIS SCI	10.1167/iovs.16-21289	284	35.5	8.56
SAMARA WA, 2017, OPHTHALMOLOGY	10.1016/j.ophtha.2016.10.008	278	34.75	8.38
JIRSA VK, 2017, NEUROIMAGE	10.1016/j.neuroimage.2016.04.049	251	31.38	7.57

publications (NP = 47) from 2015. However, SARRAF DAVID has been the influential researcher in the aspect with the highest Citations per Publications (CPP) of 69.67 reflecting sustained and broad influence across the field. In like manner, the authors including CHECCUCCI ENRICO, AMPARORE DANIELE and FIORI CRISTIAN showcase a balanced combination of high CPP with lower NP considering 2018 as the start year. This analysis suggests a strong and consistent influence in both volume as well as the quality of research. The authors including ALNAWAISEH MAGED and EDSTRÖM ERIK represent emerging contributors focusing on interdisciplinary applications in the field.

9.3.2.3 Journal Impact Analysis

Table 9.5 details the journal impact analysis by focusing on the identification of the most influential journals in the field of emerging immersive technologies in medical imaging. Journal RETINA leads with the highest h-index (22) and

Table 9.4 Overview of the author's impact analysis based on h-index concerning the emerging immersive technologies in medical imaging.

Author	h-index	g-index	m-index	TC	NP	CPP	Start year
NAVAB NASSIR	17	27	1.7	835	47	17.77	2015
SARRAF DAVID	14	15	1.556	1045	15	69.67	2016
PORPIGLIA FRANCESCO	13	16	1.857	760	16	47.5	2018
CHECCUCCI ENRICO	12	13	1.714	703	13	54.08	2018
AMPARORE DANIELE	11	12	1.571	690	12	57.5	2018
ETER NICOLE	11	15	1.222	486	15	32.4	2016
FIORI CRISTIAN	11	13	1.571	694	13	53.38	2018
JIA YALI	11	14	1.222	634	14	45.29	2016
ALNAWAISEH MAGED	10	13	1.111	443	13	34.08	2016
EDSTRÖM ERIK	10	11	1.429	385	11	35	2018

Total Citations (TC), making it a primary venue for high-impact research in this field. Furthermore, the findings reveal that the AMERICAN JOURNAL OF OPHTHALMOLOGY has the highest CPP (65.9), indicating that its publications have significant individual influence. Likewise, the journals with moderate h-index including WORLD NEUROSURGERY, BRITISH JOURNAL OF OPHTHALMOLOGY, and INVESTIGATIVE OPHTHALMOLOGY AND VISUAL SCIENCE are found to be emerging as key contributors to niche areas.

9.3.2.4 Institutional Impact Analysis

Figure 9.4 illustrates the institutional impact analysis by evaluating the contributions of leading institutions in advancing research on emerging immersive technologies in medical imaging. The UNIVERSITY OF CALIFORNIA ranks highest in total publications which reflects a strong emphasis on publishing high-quality research in diagnostic imaging. Likewise, JOHNS HOPKINS UNIVERSITY claims the second of the top-10 institutions with publications (NP = 147) maintaining a consistent publication trend over the years. The CAPITAL MEDICAL UNIVERSITY-affiliated hospitals and research centers contribute significantly to various imaging modalities and applications. CAPITAL MEDICAL UNIVERSITY stands as a leader in medical imaging research in China with its departments and affiliated hospitals actively engaged in advancing imaging techniques and applications across various medical fields. Furthermore, HARVARD MEDICAL COLLEGE has been involved in creating detailed maps of brain connections by contributing to a deeper understanding of neural circuits and potential treatments for neurological disorders. This indicates growing contributions to innovative and accessible solutions in the field.

Table 9.5 Overview of the journal impact analysis based on h-index concerning the emerging immersive technologies in medical imaging.

Source	h-index	g-index	m-index	TC	NP	CPP	Start year
RETINA	22	40	2.444	1632	40	40.8	2016
PLOS ONE	19	35	1.9	1242	46	27	2015
AMERICAN JOURNAL OF OPHTHALMOLOGY	18	20	2	1318	20	65.9	2016
INTERNATIONAL JOURNAL OF COMPUTER ASSISTED RADIOLOGY AND SURGERY	17	33	1.7	1128	51	22.12	2015
LECTURE NOTES IN COMPUTER SCIENCE (INCLUDING SUBSERIES LECTURE NOTES IN ARTIFICIAL INTELLIGENCE AND LECTURE NOTES IN BIOINFORMATICS)	15	23	1.667	903	121	7.46	2016
WORLD NEUROSURGERY	15	25	1.667	671	44	15.25	2016
BRITISH JOURNAL OF OPHTHALMOLOGY	14	19	1.556	739	19	38.89	2016
INVESTIGATIVE OPHTHALMOLOGY AND VISUAL SCIENCE	14	20	1.556	931	20	46.55	2016
EUROPEAN RADIOLOGY	13	21	1.625	459	21	21.86	2017
GRAEFE'S ARCHIVE FOR CLINICAL AND EXPERIMENTAL OPHTHALMOLOGY	12	23	1.333	649	23	28.22	2016

9.3.3 Research Focus and Emerging Topics

9.3.3.1 Keyword Co-occurrence Analysis

The keyword co-occurrence analysis has been performed using VOSviewer with an effort to construct and visualize network of keywords which results in a detailed understanding of thematic structures. The dataset comprises 1436 unique keywords which have been the author-provided keywords to analyze thematic patterns and relationships across publications in the field of emerging medical imaging technologies. Likewise, the keywords are considered co-occurring when they appear together within the same document such that the strength of co-occurrence links has been calculated based on the frequency and consistency

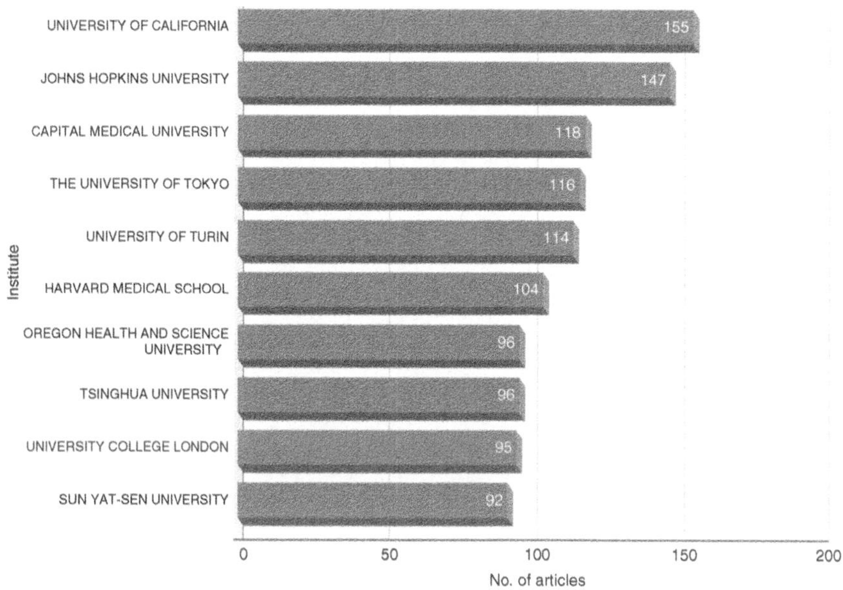

Figure 9.4 Institutional impact analysis over the last decade concerning the integration of immersive technologies with medical imaging.

of keyword pairings. Thus, 250 keywords in total have exhibited measurable total link strengths representing their interconnectedness within the research network as illustrated in Figure 9.5. Likewise, the links between nodes have indicated the presence of co-occurrence relationships alongside their thickness which provides an overview of the link strength. Thus, the red cluster with the keyword as "diagnostic imaging" highlights a dense connection with other clusters of keywords, representing areas of intense research focus. Furthermore, the new keywords associated with immersive technologies, such as "virtual reality" or "augmented reality," can be visually distinguished by their recent co-occurrence patterns with the keyword "medical imaging." Overall, the high-frequency keywords, such as "augmented reality," "virtual reality," "human/humans," "medical imaging," and "diagnostic imaging," have formed the backbone of thematic clusters. Overall, this analysis underscores the critical and emerging areas in medical imaging technologies while identifying gaps and opportunities for innovation and collaboration to drive the field forward.

9.3.3.2 Hot Topics and Emerging Trends

The word map conducted using Biblioshiny has highlighted the frequency and prominence of keywords which showcases the research community's focus and evolving interests. The Biblioshiny word map as illustrated in Figure 9.6

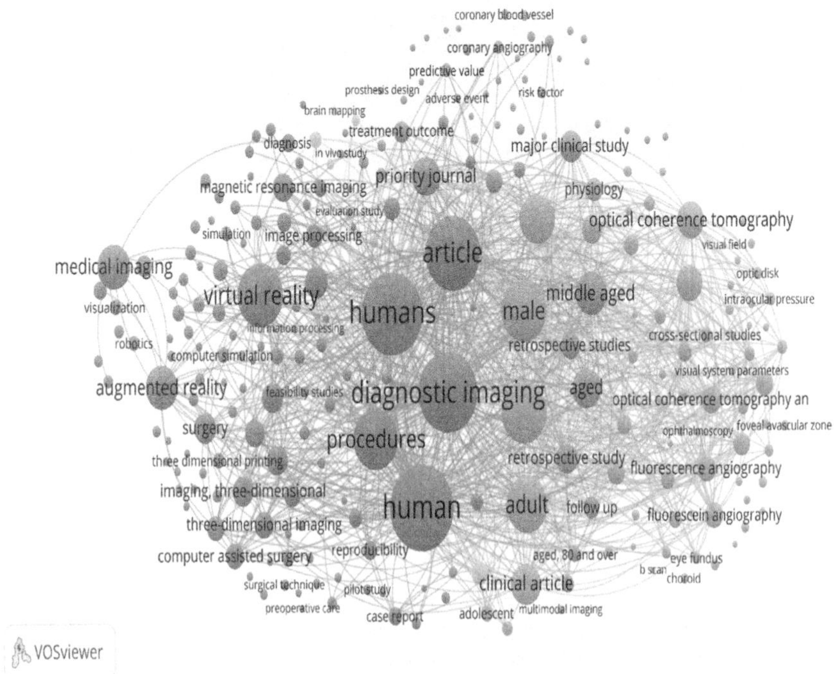

Figure 9.5 Keyword co-occurrence analysis over the last decade concerning the integration of immersive technologies with medical imaging.

Figure 9.6 Hot topics and emerging trends concerning the integration of immersive technologies with medical imaging.

has effectively highlighted hot topics like "augmented reality," "virtual reality," "diagnostic imaging," and "medical imaging" while showcasing emerging trends such as the role of various imaging techniques including "optical coherence tomography," "X-ray computed tomography," and "nuclear magnetic resonance imaging." Also, the major focus areas include the following:

- Immersive education tailored for patient profiles [33].
- AR-based anatomy learning tools [33, 35].
- Combining AR/VR for hybrid surgical navigation systems [24].
- VR systems for practicing complex surgical procedures [31].
- Multisensory XR applications for medical image analysis [37].

9.3.4 Collaboration Patterns

9.3.4.1 Author Collaboration Network Analysis

The author's collaboration network analysis explores the relationships and partnerships among researchers in the field. Figure 9.7 details the visualization for the structure of collaborations alongside the key contributors and the density of connections in the research community. The network consists of multiple clusters with different colors such that each cluster represents a group of authors who frequently collaborate. Further, the central nodes (i.e. the cluster with red color) in the network indicate highly influential authors who act as hubs for

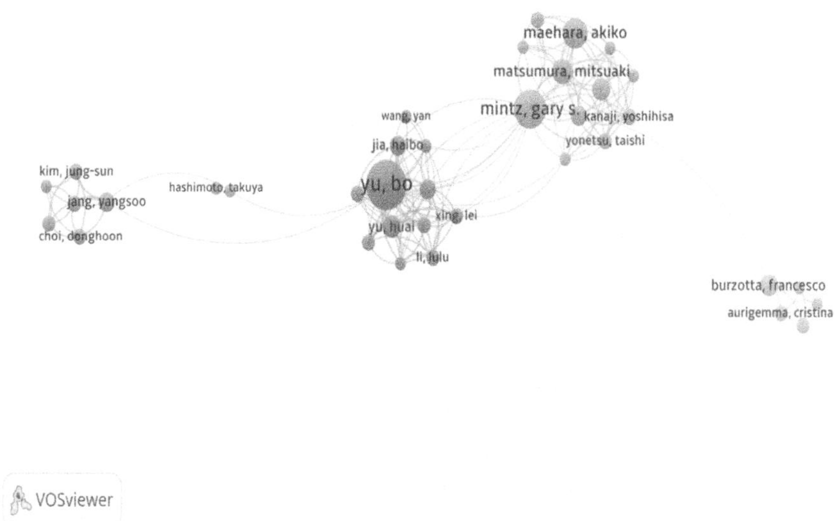

Figure 9.7 Author collaboration network analysis over the last decade concerning the integration of immersive technologies with medical imaging.

collaboration. These core authors often have a high number of co-authored publications alongside strong links across multiple clusters (i.e. the cluster with green color). Thus, the areas of high density in the network reflect active collaboration hubs, where research activity is concentrated. Overall, the network structure reveals core authors and their interdisciplinary collaborations by providing valuable insights into the collaborative dynamics driving innovation in this field.

9.3.4.2 Regional Collaboration Network Analysis

The regional collaboration network analysis as illustrated in Figure 9.8 has explored the geographical dynamics of research partnerships. The analysis has helped in identifying the key contributions of regions with cross-border collaborations and the strength of regional networks in advancing research concerning the integration of immersive technologies with medical imaging. The regions in Figure 9.8 are represented as nodes with their size indicating publication output.

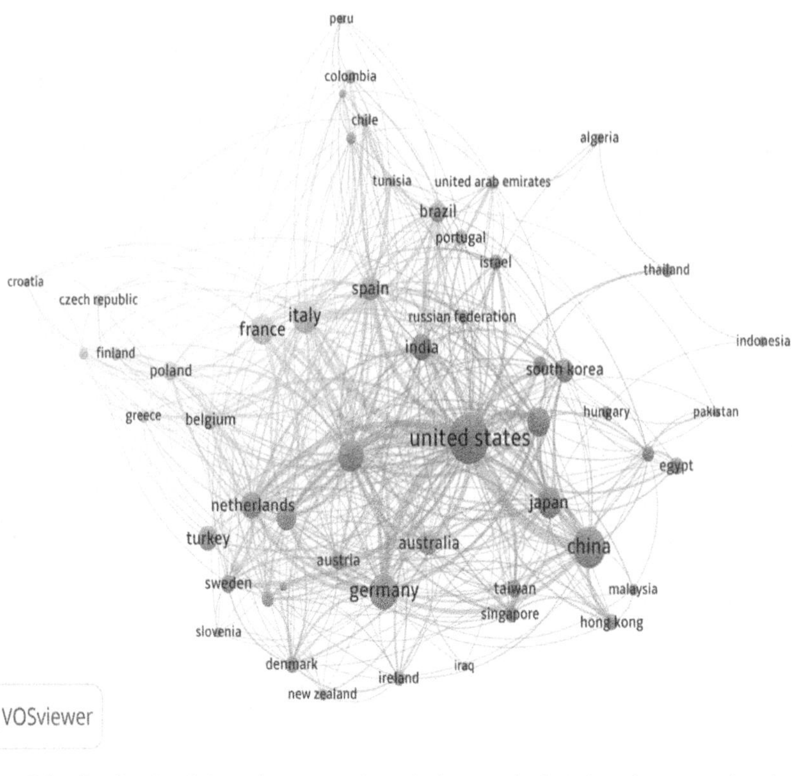

Figure 9.8 Regional collaboration network analysis over the last decade concerning the integration of immersive technologies with medical imaging.

Likewise, the edges represent collaboration links, with a thickness corresponding to the strength of partnerships. The different colored regions represent the clusters of regional collaborations with each color indicating a group of regions that frequently collaborate on specific themes or research areas. These colors have helped in identifying the dominant regions and thematic focus areas concerning the integration of immersive technologies with medical imaging. The following clusters are arranged as:

- **Red Cluster (Dominant Leaders)**: The regions include the United States, India, Japan, and China representing highly active and influential research hubs with significant inter-regional collaborations. This cluster represents the strong ties with other clusters by serving as global leaders in collaborative efforts.
- **Blue Cluster (Emerging Leaders)**: The regions including the Netherlands, Germany, and Turkey have reflected rapid growth in publication output and cross-regional partnerships.
- **Yellow Cluster (Emerging Contributors)**: The cluster includes France, Italy, and Poland which have represented the regions with growing research output and regional relevance.

9.4 Discussions and Limitations

The analysis relied solely on data from a specific database (SCOPUS) and does not include other significant repositories like Web of Science or PubMed. Likewise, the exclusion of additional databases may have limited the comprehensiveness of the analysis by overlooking relevant publications. This bias may obscure the potential of lesser-studied or resource-constrained areas. Furthermore, the analysis has been restricted to keywords explicitly listed by authors by potentially missing implicit themes or topics not captured in metadata. Thus, some of the evolving terminologies in emerging fields like XR can lead to inconsistencies in keyword classification. While co-occurrence networks provide valuable insights, the keywords with high frequency might dominate the analysis, overshadowing niche but impactful terms. Furthermore, the collaboration network has highlighted well-connected authors, but there may be potential chances of marginalizing individual researchers or smaller collaborative groups with fewer publications. Additionally, the metrics like publication count and citation strength prioritize measurable output over the practical impact of research. Thus, this quantitative analysis has offered an objective foundation which can be complemented with qualitative studies to evaluate real-world applications and innovations.

Bibliometric analysis has inherent limitations, but the analysis provides a robust framework for understanding research dynamics and identifying trends. Overall, the positive outlook lies in the continuous evolution of methodologies, tools, and collaborations that can overcome these challenges. Overall, the bibliometric analysis remain a vital instrument by embracing inclusivity, leveraging advanced technologies, and fostering global partnerships with an effort for guiding future research and innovation concerning the integration of immersive technologies with medical imaging.

9.5 Conclusion

The bibliometric analysis highlighted the emerging trends and innovations related to the medical imaging technologies, particularly in integrating immersive technologies into clinical and diagnostic workflows. The study emphasizes the growth in global research contributions alongside the leading regions including the United States dominating innovation and emerging regions like Europe and Asia showcasing increased involvement. The findings have shown the robust author and institutional collaborations alongside the regional partnerships which has played a crucial role in advancing interdisciplinary research. Further, the analysis has highlighted the identification of the most influential papers with technologies including next-generation immersive visualization, semantic segmentation, AR-based anatomical education, and the use of VR in complex surgical procedures. Thus, the strong co-authorship networks and regional clusters have underscored the importance of collaborative efforts to address global healthcare challenges. The future research should investigate adaptive AI models capable of personalizing XR-driven diagnostics and treatment plans utilizing patient-specific data. Overall, the integration of immersive technology with emerging trends in modalities of medical imaging has the potential to enhance real-time decision-making by minimizing diagnostic errors and improving patient outcomes.

References

1 Hessenbruch, A. (2002). A brief history of x-rays. *Endeavour* 26 (4): 137–141.

2 Huang, H., Zheng, Y., Chang, M. et al. (2024). Ultrasound-based micro-/nanosystems for biomedical applications. *Chemical Reviews* 124 (13): 8307–8472.

3 Katti, G., Ara, S.A., and Shireen, A. (2011). Magnetic resonance imaging (MRI)—a review. *International Journal of Dental Clinics* 3 (1): 65–70.

4 Buzug, T.M. (2011). Computed tomography. In: *Springer handbook of medical technology* (ed. R. Kramme, K.-P. Hoffmann, and R.S. Pozos), 311–342. Berlin, Heidelberg: Springer Berlin Heidelberg.

5 Pisano, E.D. and Yaffe, M.J. (2005). Digital mammography. *Radiology* 234 (2): 353–362.

6 Wester, H.J. (2007). Nuclear imaging probes: from bench to bedside. *Clinical Cancer Research* 13 (12): 3470–3481.

7 Ghosh, S., Mallick, J., Chakraborty, M., and Gupta, S. (2024). Automated bone fracture detection in X-ray imaging to improve orthopaedic diagnostics in healthcare. *Procedia Computer Science* 233: 832–840.

8 Tahir, A., Saadia, A., Khan, K. et al. (2024). Enhancing diagnosis: ensemble deep-learning model for fracture detection using X-ray images. *Clinical Radiology* 79 (11): e1394–e1402.

9 Räsänen, K., Soisalo, K., Markula-Patjas, K. et al. (2024). Increased adiposity in juvenile idiopathic arthritis—a comparison of the bioelectrical impedance analysis and dual-energy X-ray absorptiometry for the assessment of body composition. *Acta Paediatrica* 113: 2072–2080.

10 Masand, V.H., Al-Hussain, S.A., Alzahrani, A.Y. et al. (2024). Estrogen receptor alpha binders for hormone-dependent forms of breast cancer: e-QSAR and molecular docking supported by X-ray resolved structures. *ACS Omega* 9 (14): 16759–16774.

11 Yan, L., Ling, S., Mao, R. et al. (2024). A deep learning framework for identifying and segmenting three vessels in fetal heart ultrasound images. *Biomedical Engineering Online* 23 (1): 39.

12 Zheng, L., Sun, X., Liu, Y. et al. (2024). Diagnostic efficiency of ultrasound diagnosis of venous catheters and venous catheter blood flow spectrum. *Journal of Radiation Research and Applied Sciences* 17 (1): 100802.

13 Fiorentino, M.C., Villani, F.P., Di Cosmo, M. et al. (2023). A review on deep-learning algorithms for fetal ultrasound-image analysis. *Medical Image Analysis* 83: 102629.

14 Sethi, M., Ahuja, S., and Bawa, P. (2021, October). Classification of Alzheimer's disease using neuroimaging data by convolution neural network. *2021 6th International Conference on Signal Processing, Computing and Control (ISPCC)*, 402–406. IEEE.

15 Combes, A.J., Clarke, M.A., O'Grady, K.P. et al. (2022). Advanced spinal cord MRI in multiple sclerosis: current techniques and future directions. *NeuroImage: Clinical* 36: 103244.

16 Goni, M.R., Ruhaiyem, N.I.R., Mustapha, M. et al. (2022). Brain vessel segmentation using deep learning—a review. *IEEE Access* 10: 111322–111336.

17 Jaufuraully, S., Dromey, B., Story, L. et al. (2022). Magnetic resonance imaging in late pregnancy to improve labour and delivery outcomes—a systematic literature review. *BMC Pregnancy and Childbirth* 22 (1): 949.

18 Dubois, J., Alison, M., Counsell, S.J. et al. (2021). MRI of the neonatal brain: a review of methodological challenges and neuroscientific advances. *Journal of Magnetic Resonance Imaging* 53 (5): 1318–1343.

19 Garrett, H.V., Brodsky, J.E., Ahmad, T. et al. (2025). False-negative review from the mammography audit: refining breast imaging practice. *Radiographics* 45 (2): e240128.

20 Chakraborty, S., Misra, B., and Mridha, M.F. (2025). Enhancing intelligent medical imaging to revolutionize healthcare. In: *Smart Medical Imaging for Diagnosis and Treatment Planning* (ed. N. Dey, B. Misra, and S. Chakraborty), 3–20. Chapman and Hall/CRC.

21 Sarker, S., Biswas, A., Al, N.M.A. et al. (2023). Case studies on x-ray imaging, mri and nuclear imaging. In: *Data driven approaches on medical imaging* (ed. B. Zheng, S. Andrei, M.K. Sarker, and K.D. Gupta), 207–225. Springer Nature Switzerland: Cham.

22 Dukart, J., Holiga, S., Rullmann, M. et al. (2021). Juspace: a tool for spatial correlation analyses of magnetic resonance imaging data with nuclear imaging derived neurotransmitter maps. *Human Brain Mapping* 42 (3): 555–566.

23 Borrego, D. and Balter, S. (2024). A radiation dose catalog for medical imaging in pediatric procedures. *Radiology* 310 (1): e233467.

24 Kantor, T., Mahajan, P., Murthi, S. et al. (2024). Role of eXtended reality use in medical imaging interpretation for pre-surgical planning and intraoperative augmentation. *Journal of Medical Imaging* 11 (6): 062607–062607.

25 Wang, L.J., Casto, B., Reyes-Molyneux, N. et al. (2024). Smartphone-based augmented reality patient education in radiation oncology. *Technical Innovations & Patient Support in Radiation Oncology* 29: 100229.

26 Windhausen, A., Heller, J., Hilken, T. et al. (2024). Exploring the impact of augmented reality smart glasses on worker well-being in warehouse order picking. *Computers in Human Behavior* 155: 108153.

27 Rajput, M., Dudeja, R.K., and Gargrish, S. (2024). Review in healthcare using augmented reality/virtual reality: an IoT perspective. In: *Artificial intelligence technology in healthcare* (ed. N. Sharma, D. Srivastava, and D. Sinwar), 105–120. Boca Raton: CRC Press.

28 Koolivand, H., Shooreshi, M.M., Safari-Faramani, R. et al. (2024). Comparison of the effectiveness of virtual reality-based education and conventional teaching methods in dental education: a systematic review. *BMC Medical Education* 24 (1): 8.

29 Taherysayah, F., Malathouni, C., Liang, H.N., and Westermann, C. (2024). Virtual reality and electroencephalography in architectural design: a systematic review of empirical studies. *Journal of Building Engineering* 85: 108611.

30 Tanbeer, S.K. and Sykes, E.R. (2024). MiVitals–mixed reality interface for vitals monitoring: a HoloLens based prototype for healthcare practices. *Computational and Structural Biotechnology Journal* 24: 160–175.

31 Garg, D., Dubey, N., Goel, P. et al. (2024). Improvisation in spinal surgery using AR (augmented reality), MR (mixed reality), and VR (virtual reality). *Engineering Proceedings* 59 (1): 186.

32 Qu, Z., Lau, C.W., Simoff, S.J. et al. (2022). Review of innovative immersive technologies for healthcare applications. *Innovations in Digital Health, Diagnostics, and Biomarkers* 2 (2022): 27–39.

33 Tricomi, P.P., Nenna, F., Pajola, L. et al. (2023). You can't hide behind your headset: user profiling in augmented and virtual reality. *IEEE Access* 11: 9859–9875.

34 Ryan, G.V., Callaghan, S., Rafferty, A. et al. (2022). Learning outcomes of immersive technologies in health care student education: systematic review of the literature. *Journal of Medical Internet Research* 24 (2): e30082.

35 van der Kruk, S.R., Zielinski, R., MacDougall, H. et al. (2022). Virtual reality as a patient education tool in healthcare: a scoping review. *Patient Education and Counseling* 105 (7): 1928–1942.

36 Innocente, C., Ulrich, L., Moos, S., and Vezzetti, E. (2022). Augmented reality: mapping methods and tools for enhancing the human role in healthcare HMI. *Applied Sciences* 12 (9): 4295.

37 Krieger, K., Egger, J., Kleesiek, J. et al. (2025). Multisensory extended reality applications offer benefits for volumetric biomedical image analysis in research and medicine. *Journal of Imaging Informatics in Medicine* 38: 646–655.

38 Arruda, H., Silva, E.R., Lessa, M. et al. (2022). VOSviewer and bibliometrix. *Journal of the Medical Library Association: JMLA* 110 (3): 392–395.

39 Pettersen, E.F., Goddard, T.D., Huang, C.C. et al. (2021). UCSF ChimeraX: structure visualization for researchers, educators, and developers. *Protein Science* 30 (1): 70–82.

40 Minaee, S., Boykov, Y., Porikli, F. et al. (2021). Image segmentation using deep learning: a survey. *IEEE Transactions on Pattern Analysis and Machine Intelligence* 44 (7): 3523–3542.

41 Olarte, O.E., Andilla, J., Gualda, E.J., and Loza-Alvarez, P. (2018). Light-sheet microscopy: a tutorial. *Advances in Optics and Photonics* 10 (1): 111–179.

42 Mo, Y., Wu, Y., Yang, X. et al. (2022). Review the state-of-the-art technologies of semantic segmentation based on deep learning. *Neurocomputing* 493: 626–646.

43 Jirsa, V.K., Proix, T., Perdikis, D. et al. (2017). The virtual epileptic patient: individualized whole-brain models of epilepsy spread. *NeuroImage* 145: 377–388.

10

Therapeutic Interventions Guided by Advanced Imaging Modalities

Manisha Pathania and Chander Partap Singh

University Institute of Computing, Chandigarh University, Mohali, Punjab, India

10.1 Introduction

The amalgamation of technology and education has changed the way we learn and share information. In medical education, traditional methods using lectures, textbooks and direct supervision often do not achieve a suitable theoretical fundamentals/hands-off expertise interface. Teachers, clinicians, and researchers are looking for new and exciting methods and tools to make learning more effective, enhance precision and efficiency in patient care. Augmented reality (AR) is one of these technologies that is considered revolutionary, as it provides immersive, interactive, and engaging 3D environments for medical training [1–4].

In the field of medicine and therapeutic settings, students and medical professionals require a deep knowledge of anatomy, complicated procedure development, and decision-making related to patient care. They require both mental and motor skills. Conventional teaching methods, while useful to a limited level, face restrictions like the high cost of specialized training equipment, ethical issues, and limited access to real patients [5]. Cadaver-based learning which is supposed to be the gold standard of anatomical learning has many challenges such as preservation and cultural issues. In the same way, training on live patients can pose a risk to the safety of patients, especially if novices learn [6]. Due to these limitations, there is an urgent need for alternative measures to conduct educationally serious activities in an accessible manner and with safe locusts.

AR, which shows information on the world around you, is positioned to help solve these problems. AR enables learners to engage with virtual elements in a real-world context, providing an interactive experience with digital represen-

Biomedical Imaging Technology: Signal Processing Strategies and Innovations, First Edition.
Edited by Ayush Dogra, Shalli Rani, and Ankita Sharma.

tations of anatomical structures, clinical scenarios, or procedural simulations. This capability not only improves spatial understanding but also offers a safe platform for unlimited practice, an essential step for mastering complex medical operations [7–9].

10.1.1 The Role of AR in Medical Training and Imaging-Guided Therapeutic Interventions

AR is popular in medical education as it helps close the disparity between abstract theoretical knowledge and clinical practice. To illustrate, the Microsoft HoloLens AR application, for example, allows students to superimpose a 3D model of the anatomy onto a physical space, offering more interaction than static pictures in books [10]. These tools foster a greater understanding of spatial relations within the human body, which is crucial for surgery and radiology [11]. In surgical training, AR systems like ImmersiveTouch and Osso VR give guidance and feedback when performing procedures, which the trainees perform with a high degree of accuracy [12]. Research shows that such platforms help to enhance procedural competency while using less time for skill acquisition [13]. Furthermore, AR assists in high-risk scenarios such as neurosurgical planning and orthopedic procedures by integrating imaging data with real-time visualization, thereby reducing error rates and improving patient safety.

AR is much more than honing skills. In emergency response training, for example, AR-based simulations duplicate high-stress situations. This enables learners to practice important skills in lifelike environments. These simulations help train healthcare professionals to perform optimally in rare events like mass casualty incidents or intricate cardiac procedures [14–16]. When we use AR in learning, it brings more realism and interactivity. AR also provides immediate feedback.

10.1.2 Educational Theories Underpinning AR Use

The strength of AR pedagogy for medical training is grounded in educational theories with prior evidence using experiential learning, constructivism, and situated cognition to support the assertion that effective medical training via immersive virtual reality (VR) enhances active learner engagement and learning. Experiential learning theory advocates knowledge construction through hands-on, reflective experiences. AR offers a platform for active experiments, in which learners can operate on digitally represented objects, run experiments, and observe the consequences in real time [17]. Constructivist approaches to education suggest that knowledge is more effectively gained when students work with their environment. Students can explore complex medical concepts, largely intentional in AR, and this is supported by AR's interactive interfaces that facilitate individualized learning [18].

Learning in the situated view is most effective when embedded in the environment where the learning will be implemented. By simulating real-world medical scenarios, AR simulations replicate and help develop skills that are transferrable to policy practice. For example, an AR-based catheter insertion simulation instructs and places this inpatient care context [19]. By simulating therapeutic interventions, AR ensures that medical professionals develop skills transferrable to clinical practice.

10.1.3 Evidence of AR's Impact on Learning Outcomes

Field research has shown that AR is an effective way to improve learning outcomes of medical education in all domains specially in therapeutic procedures. Woodall et al. [20] performed a systematic review of the literature and found that AR significantly increased knowledge retention learner satisfaction and procedural accuracy compared to traditional methods. A second study by Küçük et al. [21] showed that AR-based anatomy education enhanced spatial cognition and information recall by medical students compared to conventional textbook-based learning. AR systems are also found similarly to reduce error rates and strengthen procedural confidence in the surgical learning process, especially for novice learners [22].

AR has taken one notable example, where AR overlays are used in laparoscopic training to help trainees navigate surgical instruments in a virtual abdomen. This approach not only enhances technical skills development but also teaches about spatial orientation and decision-making under pressure [23]. Furthermore, AR is instrumental in closing medical education gaps by providing scalable and inexpensive training tools for resource-constrained environments [24].

10.1.4 Evidence of AR's Impact on Learning Outcomes

The integration of AR in medical training syncs with the objective of current-day education, which focuses on accessibility, equity, and sustainability. AR democratizes access to high-quality training resources to remedy longstanding disparities in medical education. For example, AR applications can be used by institutions in low-resource settings to perform complex simulations of surgical procedures that do not require expensive equipment or cadaveric specimens [25].

AR also allows lifelong learning, an essential aspect of healthcare professionalism. The rapid development of medical knowledge and technology forces medical practitioners to keep on improving their skills. Ongoing education made available using AR platforms provides ample flexibility and engagement to support practitioners to stay up to date without interfering with professional responsibilities [26].

10.1.5 Future Perspectives

AR has enormous potential in medical education and therapeutic interventions, which is barely explored. AR applications are expected to be enhanced by emerging technologies such as artificial intelligence (AI) and machine learning which will enable adaptive learning experiences appropriate to specific needs. For example, AI algorithms could examine how a student performs in real time and offer personalized feedback and suggestions for improvement. Just like in wearable devices and sensors, AR can integrate with them to create real-time physiological monitoring during simulation to get an approximation of stress levels, decision-making patterns, and motor skills [27–29].

Despite these challenges, the realization of this potential depends upon successfully addressing high development costs, technical barriers, and the demand for robust evidence regarding the efficacy of AR. Through partnerships with educators, technologists, and policymakers, AR must be used sufficiently in medical curricula and available to all learners.

The exploration of AR in medical training in this comprehensive way recognizes the transformative potential of this gap-bridging technology. Rejuvenating medical education through the use of AR yields immersive, interactive, and scalable learning environments addressing critical limitations of currently available methods allowing for an equitable future in which medical education can be more effective and engaging.

10.2 Background and Significance

10.2.1 The Current Landscape of Medical Education

Medical education is a very demanding field that asks students and professionals to learn not only the theory but also practical, decision-making and motor skills under high-pressure environments. Classroom lectures, cadaver-based anatomy teaching, and supervised clinical practice have all been traditional methods used in training generations of providers. Despite being limited, these approaches are incapable of handling the growing complexity of medical science and the demand for hands-on, experiential learning [1–3].

For instance, the leading example of anatomical education—cadaveric dissection has challenges on both ethical and pragmatic grounds: high costs, logistical problems to prepare and maintain specimens, etc. [4, 5]. Similarly, while very valuable, supervised clinical training is also constrained by limited access to diverse cases, high patient loads, and the need to balance educational opportunities with patient safety [6]. These constraints make it clear that innovative tools to complement conventional methods are needed, which should offer scalable, flexible, coherent, and engaging learning experiences—any one of which should not jeopardize educational rigor or the health of patients.

10.2.2 Augmented Reality as a Solution

AR has evolved into a clinical revolution, addressing many of the challenges of medical education. By AR, digital content can be superimposed over the physical environment to interact with virtual representations of anatomical structures, surgical cases, and clinical environments in real time [7, 8]. In contrast with traditional techniques, AR allows for the practicing of complex operations repeatedly, in a controlled risk-free environment. It is particularly easy to use in situations where even the smallest errors can have critical results like in surgery.

The major advantage of AR is the ability to provide spatial and procedural understanding. For example, AR-based anatomy apps like the HoloAnatomy app enable students to view and manipulate 3D models of the human body to achieve a better understanding of spatial relations between organs and systems' functional interconnections. In addition to improving knowledge retention, these tools also bridge the gap between theoretical learning and clinical application, a gap that is usually absent in a traditional curriculum [10].

10.2.3 Historical Development of AR in Medical Training

AR application in medical training goes back to the technological improvements in visualization and simulation. An early focus was on VR surgical training environments, providing proof of concept for the use of immersive simulations to improve procedural skills [10]. Yet VR was often tethered to isolated setups and specialized hardware, and as such, was not very accessible or scalable. However, with AR's capacity to merge virtual content into real environments through devices such as smartphones, tablets, and head-mounted displays (HMDs), it quickly became a compelling proposition that's more readily available [12, 13].

Among significant milestones in AR development are AR-assisted surgical simulators such as ImmersiveTouch and Osso VR which allow trainees to have realistic haptic feedback and step-by-step guidance for simulated procedures [14]. Thanks to these platforms, surgical education has progressed significantly, and users can practice skills like laparoscopic navigation, sutures, and organ resections with high fidelity and precision [15]. Previous studies have demonstrated in AR-based training that accuracy improves and error rates reduce in particular for novices [16, 17].

10.2.4 Addressing the Gap Between Theory and Practice

One of the most valuable gives of AR to medical education is its ability to bridge the space between the theory and its application. However, the vast majority of traditional teaching methods still involve a significant amount of didactic instruction, where students might learn abstract concepts through lectures or books. This approach, although useful for acquiring foundational knowledge, is lacking for learning to endure the dynamic and unpredicted nature of real-world medical practice [18]. AR fills this limitation, contextualizing theoretical knowledge within interactive and immersive environments where concepts can be applied in simulated clinical scenarios.

AR-based emergency response simulations, e.g. cardiac arrests, trauma resuscitations, and mass casualty injuries, help one replicate high-pressure events. These simulations provide learners with the opportunity to practice critical decision, team, and technical skill development in a realistic setting (controlled environment) [19]. Such training has been found to improve both technical proficiency and confidence and resilience, important qualities in healthcare professionals [20, 21].

10.2.5 Accessibility and Scalability

AR also has the potential to facilitate the democratization of access to quality medical education. Often these traditional resources are prohibitively expensive and inaccessible to institutions in low-resource settings such as cadavers, advanced imaging systems, and surgical simulators [22]. AR, however, is better suited to deployment on more accessible devices, such as smartphones and tablets, as they offer less expensive options. AR is used by platforms such as Anatomy 4D and Complete Anatomy to deliver interactive anatomical content to students and educators to have access to advanced learning resources regardless of geographic or financial constraints [23, 24].

10.2.6 Enhancing Learner Engagement and Retention

Education theories like experiential learning and constructivism stress active participation and hands-on experiences in learning knowledge. First, it is consistent with these principles in that the AR interface promotes exploration, experimentation, and self-directed learning [25]. It has been shown that student engagement, motivation, and satisfaction are greater in AR-based learning than in more traditional methods [26, 27]. AR also can present information in multimodal format (visual, auditory, and tactile) which enhances cognitive retention and reduces cognitive load, thus enabling better learning [28].

10.2.7 Challenges in AR Adoption

AR has many advantages, but it also has its challenges in being adopted for medical education. There are high development costs, technical barriers, and the need for specialized training for educators and students. In addition, AR works well only if its content is of high value and relevant enough. Improperly designed AR applications can result in misinformation, low user engagement, actually cognitive overload [30]. These challenges can only be addressed by partnerships among educators, technologists, and policymakers to make AR tools easy to use, based on evidence and aligned with curriculum objectives.

10.3 Core Applications of AR in Medical Training

AR has transformed the medical learning curve by linking theory to practice through the lens of dynamic and immersive experiences that allow students as well as practitioners to immerse themselves in the learning process. AR has brought forth new ways for visualization, simulation, and interactivity of medical education through overlaying of digital information over the real-world context. We elaborate on how AR core applications can shift the status quo in the domains of medical training.

10.3.1 Anatomy Education

Anatomy education is one of the most pervasive applications of AR in medical training, where AR augments traditional ways of learning human anatomy. However, cadaver dissection, a key component of anatomy education, faces ethical concerns, and preservation hurdles, and is costly. However, the AR offers an equally compelling alternative by offering interactive, three-dimensional (3D) human anatomy visualizations that students can explore from multiple perspectives [3, 4].

There are applications such as "HoloAnatomy" and "Complete Anatomy" that allow users to detail examine organs, tissues, and systems. This deepens learners' understanding by providing interaction with virtual structures that they can rotate and cross-sectional layers, specifically where complex spatial relationships are involved. Studies indicate that students who use AR-based tools show better knowledge retention and spatial cognition compared to traditional learning [5, 12].

In addition, AR enables collaborative learning. For example, an anatomy 4D platform allows groups of students to engage with the same AR model, all shared simultaneously leading to discussion and shared understanding [14]. However, in classrooms or remote learning where working with physical specimens may be prevented by space constraints or distance, this is particularly helpful.

10.3.2 Surgical Training

While traditional apprenticeship models have been challenged by surgical training with AR to overcome limited OR time and patient safety concerns, recent work has shifted focus toward demonstrating AR's relevance for teaching more advanced techniques. Simulators for AR can give trainees risk-free environments to practice and sharpen their skills before they work on real patients. ImmersiveTouch and Osso VR are just a couple of systems that provide a simulation as realistic as surgery, with haptic feedback to replicate the sensations of surgery [15].

An important advantage of AR in surgical training is that it can provide step-by-step guidance. For instance, AR headsets, such as Microsoft HoloLens, can project virtual overlays onto surgical fields, so trainees can see them complete procedures as they go in real time. Not only does this help skill acquisition but also improves the performance of complex surgeries [13, 22]. According to research, AR-based training offers an improvement in procedural accuracy and error rates, most notably in novice surgeons [11].

Preoperative planning and intraoperative navigation are other ways that AR is used. AR superimposes 3D models of a patient's anatomy on their body to help surgeons visualize under-the-skin structures like tumors or blood vessels, reducing risk and improving precision in surgery [23]. In particular, this application has been valuable for minimally invasive and robotic-assisted surgeries [18].

10.3.3 Emergency Medicine and Trauma Training

In the realm of emergency medicine, where rapid, potentially high-pressure decisions need to be made, AR has been well served. AR-based simulations can reproduce real-life emergency scenarios, for example, cardiac arrests, trauma resuscitations, and disaster responses offer condition training in controlled settings for learners to practice teamwork and exact skills [21, 28].

Readily available AR platforms, such as "CardioSim," simulate cardiac emergencies as well as practice lifesaving procedures such as defibrillation and intubation. These simulations use real-time physiological feedback to allow learners to evaluate the results of their actions and adjust accordingly [16]. AR-based emergency training, while studies show, improves technical proficiency and non-technical skills, such as communication or leadership, that are key to situations in high-stress environments.

AR is used in trauma training to visualize internal injuries and to guide interventions. For instance, AR could superimpose images of fractures or internal bleeding onto a patient, giving trainees a clear mental model of how those injuries happened and how they should be treated [30].

10.3.4 Medical Imaging and Diagnostics

By adding imaging data, such as CT scans and MRIs, into augmented views of patients, AR enhances medical imaging and diagnostics. Enabling trainees to view their internal anatomy in real time, helps a trainee to understand how diagnostics are performed. For example, AR applications in radiology allow users to use mannequins or even real people to overlay diagnostic images on top of them, thereby allowing hands-on interpretation of scans [31].

In interventional radiology, AR is utilized to guide needle placement during biopsies or catheter insertion. AR sends real-time feedback by projecting imaging data onto the patient's body and using that to project the patient into the system, enabling better accuracy and a decreased risk of complications [19]. In particular, these capabilities are useful in training where learners can practice complex procedures without putting patients at risk.

10.3.5 Procedural Simulations and Skill Training

VR and AR are applied almost universally to basic training in routine medical procedures, such as venipuncture, catheterization, and suturing. For example, AR is used by applications like AccuVein to project vein images onto the skin so that trainees can locate veins accurately during blood draws or IV placements [32]. Such tools not only increase procedural accuracy but also decrease patient discomfort important in clinical settings [20].

AR platforms combine visual and haptic feedback for realistic surgical skill training. AR-based laparoscopic simulators are examples of the kind of machine where trainees can develop hand–eye coordination and fine motor skills while performing virtual surgery on virtual models [29]. Available research indicates that learners trained using AR simulators learn faster, develop greater skill acquisition, and retain that skill acquisition better than subjects trained using traditional methods [6].

10.3.6 Patient Communication and Empathy Training

Like communication and empathy in medical training, AR is also used to improve communication and empathy in medical education. AR applications include simulating patient experiences, for example, to enable learners to understand patients' perspectives of, for example, individuals with dementia or chronic pain. Trainees create a more nuanced awareness of the consequences of illness by using AR to experience visual impairments or physical limitations [27].

Empathy training utilizing AR has been shown to result in improved patient–provider interaction[s] and increased patient satisfaction [26]. AR is also used to teach "effective communication skills" including telling someone something bad, using a realistic patient simulation [25].

10.3.7 Remote and Collaborative Learning

The COVID-19 pandemic has brought to the forefront remote learning tools used in medical education. But without AR enabling remote training, remote training simply wouldn't have been possible. Students and instructors can even interact with shared AR models in real time even when they are physically distant [24].

In telemedicine training, for example, AR has also been used, providing the learner with an opportunity to simulate remote patient consultations and diagnosis. By incorporating AR into telemedicine platforms, trainees can visualize patient anatomy and simulate physical examinations before they are encountered in real-world telehealth scenarios.

10.4 Bridging Theory and Practice

The application of AR in medical training addresses a long-standing challenge in medical education: the disconnect between theory and practice. Up until now, medical education has largely relied on lectures, textbooks, and sometimes even hands-on practice, meaning that learners are often not prepared for the real-world clinical environment. AR is taking the lead in filling that gap, promising immersive, interactive, and practice-oriented learning experiences. In this section, we discuss how AR closes the gap between theoretical learning and clinical application to develop the best preparation and competence among medical professionals.

10.4.1 Challenges in Traditional Medical Education

To date, medical education has historically proceeded in a stepwise fashion way, first from the foundational sciences to stellar training. This approach provides a solid theoretical basis for many students but does not usually do a sufficiently good job of preparing them for the intricacies of daily life in real-world medical practice. Studies have highlighted several limitations:

Insufficient Hands-On Practice: In resource-constrained settings, experiential learning opportunities (such as cadaveric or biological studies, patients, and simulation labs) are limited by restricted access to these facilities [3, 4].

The gap in Clinical Readiness: One of the many issues many students face when transitioning from theory to practice is that they have little experience with real clinical scenarios during training [12].

Safety Concerns: Real patients carry risks with invasive procedures, so to practice on them has risks. That makes educators a little scared to give trainees the opportunity to gain enough practical exposure [14].

By virtue of AR, the risk is removed, scalability is achieved, and the clinical setting is emulated by immersive training environments.

10.4.2 AR as a Link Between Theory and Application

AR combined with physicality provides a means to get from theoretical understanding to practical application, bridging digital and physical. Both types of learning are provided by such a dual engagement, and this is why it is effective in medical education.

Interactive Learning Environments: AR-based applications such as "HoloAnatomy" and "Visible Body" allow students to investigate anatomical structures interactively and reinforce the theory using visual and spatial understanding [13, 33]. With these platforms, learners can tear apart virtual models, manipulate organs, and grasp relationships that seem complex in a 3D space which is difficult to do in textbooks or static images [21].

Simulation of Clinical Scenarios: This AR recreates realistic medical scenarios where trainees could practice their diagnostic reasoning, clinical decision-making, and procedural skills in a controlled environment. For example, AR systems use cardiac arrest scenarios to teach advanced cardiac life support (ACLS) using real-time feedback on a learner's actions [17, 30].

Real-Time Feedback and Adaptation: AR's greatest benefit is also one of its most unique assets—the ability to provide instant feedback and help users perfect their skills. AR-assisted surgical training platforms evaluate hand movements, precision, and decision making, yielding insights that are vital to improving AR's application and software as a surgical working platform [16].

10.4.3 Enhancing Skill Acquisition and Retention

The research has shown that AR helped significantly improve skill acquisition and retention. A study conducted by comparing traditional and AR-based anatomy education found that students trained using AR displayed a 30% greater retention of anatomical knowledge after six months [11]. AR has led to faster mastery of complex procedures in surgical training, which has shown similar outcomes [18].

The reasons for AR's effectiveness in skill acquisition include:

Active Engagement: Unlike passive learning methods, AR acts as an active engagement device for learners, forcing them to interact with virtual objects and make decisions, which enhances understanding [22].

Repetition Without Risk: AR environments enable trainees to repeatedly practice procedures endlessly without fear of injuring patients, which is most useful for high-risk interventions such as making a central line or doing a laparoscopic procedure [32].

Multisensory Learning: Included in all simultaneous stimulation are the visual, auditory, and kinesthetic senses, and all are catered to in order to catch the diverse learning preferences and bolster memory retention [20].

10.4.4 Bridging Cognitive and Procedural Learning

In addition to enhancing cognitive understanding, AR closes the gap between cognitive and procedural learning. Theoretical knowledge is insufficient in the fields of surgery and emergency medicine without procedural competence. AR facilitates this integration in several ways:

Step-by-Step Guidance: Procedural instructions are overlaid onto the learner's field of view in AR, thus ensuring the correct steps are followed during a task. In procedures like intubation and suturing, this has been particularly effective [2, 26].

Simulation of Rare Scenarios: AR allows trainees to get exposed to rare or complex cases like congenital anomalies or multi-system traumas that they could not experience during traditional training [29].

Integration with Diagnostic Tools: [27] AR systems can be employed such that they blend with imaging modalities including CT or MRI, allowing learners to view the internal anatomy in real time and apply relevant theoretical knowledge to diagnostic decision-making.

10.4.5 Collaboration and Remote Learning

AR also provides another avenue through which theory and practice merge: in the form of collaborative and remote learning. In this era of AR-enabled telepresence, instructors can remotely lead learners through the clinical procedures, in real-time, with real-time feedback and support. During the COVID-19 pandemic, this capability was invaluable due to restrictions on in-person training opportunities [23, 28].

In fact, AR platforms like "Proximie" allow surgeons to extrude virtual annotations onto live video feeds that educate trainees, such as those who are in different locations, through surgical procedures. Besides democratizing access to expert knowledge, it also promotes theory into practice [34].

10.4.6 Industry and Academic Partnerships

Collaborations between academic and medical institutions and between technology and medical companies are helping to foster the segment's growth through more widespread AR use in medical training. Developed through these partnerships, these cutting-edge AR tools have been designed specifically for a particular medical specialty. For example:

AR in Orthopedics: Both training and patient outcomes are improved by companies such as AccuVein and Augmedix who have created AR systems for vein visualization and orthopedic surgery planning [35].

AR in Radiology: AR is being used by academic centers to train radiologists in interpreting complex imaging studies, thus bridging the gap between the analysis of an image and making a clinical decision [25].

Collaborating such that AR tools are evidence-based, user-friendly, and meet the needs of medical educators and physicians is essential.

10.4.7 Future Directions

With further development of AR technology, it will not only increase in potential to bridge theory and practice but also increase the speed at which it does so. Future developments include:

Integration with Artificial Intelligence (AI): AI-driven AR systems could personalize training through performance analysis and provide customized feedback [31].

Wearable AR Devices: AR's real commercial success will likely be made possible through the development of lightweight, ergonomic AR headsets and smart glasses that make AR more accessible for daily use in medical training [24].

Interdisciplinary Applications: It is predicted that AR will be a vital part of interdisciplinary training, allowing medical professionals to work with engineers, computer scientists, etc., in difficult healthcare situations [9].

10.5 Challenges and Limitations

AR is a transformative adjacent technology with great potential to enhance medical training. Nevertheless, adoption and integration are both complex and constrained by numerous issues. These addresses are rather multifaceted: they pose technology, pedagogy, financial, ethical, and accessibility concerns.

10.5.1 Technological Constraints

10.5.1.1 Hardware Limitations

Most of the AR systems rely on special hardware such as HMDs, smartphones, and tablets. While advanced HMDs, like Microsoft HoloLens or Magic Leap, offer immersive experiences, they are costly and not widely available in resource-constrained settings. Additionally, long-term use of these devices is also uncomfortable as these devices are weighed and yet the form factor is not comfortable to use for extended periods of intensive medical training sessions [3, 4, 36]. Another major issue is battery life especially when long-duration simulations of, e.g. surgical procedures or anatomy dissection are required [37].

10.5.1.2 Software Challenges

Medical training AR software needs to be reliable and effective, and this task is both complex and resource-intensive. The cost of customization is also high and the development time is long as it is needed to best address diverse medical specialties. The adoption problem is exacerbated by compatibility issues between hardware and software as well as limited cross-platform support. However, existing solutions have difficulties with achieving high fidelity and real-time responsiveness for realistic simulations [38, 39].

10.5.1.3 Latency and Real-Time Interactivity

Real-time interactions are critical in surgical training and precision fields, and latency in AR systems can break this, rendering them unusable for real-time precision applications. If no bigger delays can occur, the virtual overlays will misalign with the physical ones, thereby causing inaccuracies in the learning results. A simple miscalibration of the virtual instruments can also significantly decrease skill development during minimally invasive surgery training, for example [40]. Effective learning requires real-time interaction and still is a major technical challenge.

10.5.2 Pedagogical and Integration Challenges

10.5.2.1 Lack of Faculty Training

This means that AR integration requires educators to know the technology, be great at it, and know its pedagogical value. However, there is little training for medical educators in AR systems, and faculty tend to resist new methods when they are used to traditional teaching methods. However, the gap is often filled, because institutions do not provide structured training programs to address this problem [12, 41].

10.5.2.2 Curriculum Design and Overcrowding

Another challenge is to integrate AR into already-dense medical curricula. AR can't simply replace traditional teaching methods but instead has to supplement them. The restructuring of the curriculum becomes complicated because the curriculum must balance new AR-based learning modules such as ICBE and conventional methods including cadaver dissections, textbook study, and clinical rotations [42, 43].

10.5.2.3 Learning Curve for Students

There is tremendous variation in both the amount of technological literacy and adaptability to immersive learning environments of medical students. AR interfaces are not an option for some and cognitive overload with simulations complicates matters for others. This leads to inconsistent learning outcomes [44, 45] as students' readiness to adopt AR technology is not a constant variable.

10.5.3 Financial and Accessibility Barriers

10.5.3.1 High Costs of Implementation

What's even more expensive is the financial burden of AR technologies. Institutions in developing countries often have prohibitive costs of acquisition of hardware, software development or purchase, and training of employees. For example, large-scale deployment of HoloLens is impractical due to its high cost (several thousand dollars), making it an advanced AR system [46, 47].

10.5.3.2 Maintenance and Updates

Regularly, AR systems need an update to work effectively and securely. The financial burden includes also maintenance costs, composed of software licensing and replacement of hardware. Further, these recurring expenses may be difficult to justify institutions when budgets are limited or priorities exist [48].

10.5.3.3 Disparities in Accessibility

Leadership of high-performance AR tools can be led by advanced institutions in developed countries, but in low-resource settings, schools can't. As this inequality occurs, medical education has the effect of creating a digital divide between students in wealthier institutions, which enjoy better training resources, and their peers in underserved regions [49].

10.5.4 Ethical and Regulatory Concerns

10.5.4.1 Data Privacy and Security

Most AR systems expose sensitive data such as user performance measures and, in some cases, biometrics. The importance of ensuring compliance with data protection regulations such as HIPAA in the United States, or GDPR in Europe to protect both students and patient information. If this data cannot be secured, confidentiality breaches and the trust in AR technologies [50, 51] are in jeopardy.

10.5.4.2 Simulation Limitations

The greatest strength of AR is that it offers a risk-free environment to train medical procedures, but it cannot fully approximate the complexity of medical scenarios in real life. It is difficult to simulate accurate variations of patient anatomy, physiologic responses, and emotional contexts, which may leave students undertrained to meet real-life challenges [52].

10.5.4.3 Ethical Use of Patient Data

For certain AR applications, the use of real patient data for creating authentic simulations is required. Consider conducting such data with the ethical implications of using these data, especially in training environments, to prevent data from being used, or shared without consent [53].

10.5.5 Standardization and Accreditation Issues

10.5.5.1 Lack of Standardized Guidelines

Without standardized protocols for AR implementation in medical education, this implementation is not uniform and therefore not effective. As of yet, accrediting bodies have not established robust criteria for assessing AR-based training programs to result in inconsistencies between institutions [54, 55].

10.5.5.2 Efficacy Validation

AR's efficacy over the long term at improving learning outcomes and clinical skills is still not studied enough. Preliminary studies all correlate with promising results; however, more comprehensive multi-institutional research on the AR educational tool must be completed to prove it as a dependable method of education [56].

10.5.6 Cultural and Psychological Barriers

10.5.6.1 Resistance to Change

The adoption of AR can be hindered by cultural resistance to adopting new technologies. Faculty and administrators may be untrusting of AR as something new, and students who are not versed in unfamiliar learning modes may be unwilling to make the change [57].

10.5.6.2 Cognitive Overload

AR can enhance engagement but sophisticated or poorly designed simulations can either overwhelm users or inhibit effective training. Scenarios involving high-stakes procedures and dense information create a cognitively overloaded environment for the learners [58].

10.6 Future Directions

AR in medical training is poised for a revolution in educational methodology, closing skills acquisition gaps in an increasingly complex landscape, and democratizing access to paradigm-shifting tools. But, its full potential cannot be reached without overcoming its present limitations, creating cross-disciplinary collaborations and investing in research and development (R&D). What we present below are potential trajectories for the field, showcasing opportunities to innovate and move towards broader adoption.

10.6.1 Industry and Academic Partnerships

10.6.1.1 Improved Hardware Design

Future AR hardware, which needs to work without discomfort, will have to be lighter and more ergonomic. Challenges in prolonged training sessions can be addressed by the development of compact, wireless HMDs with extended battery life [1, 2]. Other innovations such as foldable AR glasses or an AR display based on contact lenses could change the user experience by combining usability and functionality [3].

10.6.1.2 AI-Integrated AR Platforms

AI integration into the AR systems appears, and it's promising. AI can help use simulations for adaptive learning, to personalize the simulation in response to the learner's pace and skill level [59]. For example, real-time feedback to surgical trainees leads to improvement in precision and decision-making skills [4, 5]. Not only that, AI-driven natural language processing can also help with multilingual AR interfaces and make the technology more accessible globally [6].

10.6.1.3 Interoperability Standards

The integration of AR applications across all platforms becomes seamless using universal software standards. The collaboration between technology providers, medical institutions, and entities responsible for regulations can be facilitated by standardized frameworks that promote common user experience and contain the costs for deployment of proprietary solutions [7, 8].

10.6.2 Personalized and Collaborative Learning

10.6.2.1 Customized Training Modules

Future AR systems should offer hyper-personalized training programs. These programs can examine a learner's individual learning patterns and adapt simulations to concentrate on the areas on which a trainee should focus. For instance, AR could enable a medical student to practice suturing with very difficult suturing to their ability to perform [9].

10.6.2.2 Collaborative and Remote Learning

With its ability to allow for shared virtual environments, AR has the power to completely transform group learning. For real-time interaction in a virtual operating room or anatomy lab, students and educators from different geographical locations can interact. For global medical education initiatives, collaboration across borders can make far better learning outcomes [10, 11], and this capability will matter most in that respect.

10.6.2.3 Integration with Telemedicine

AR can also support telemedicine, by supplying visual overlays when performing remote consultations or diagnostics. For instance, a remote specialist could steer a student through a diagnostic procedure with augmented visual aids to facilitate hands-on learning while physically distant [12].

10.6.3 Expanding Access and Equity

10.6.3.1 Affordable AR Solutions

"What matters is that it doesn't become too expensive." However financial barriers driven by the costs of current AR devices and proprietary software platforms can be greatly reduced through efforts to develop low-cost AR devices and open-source software platforms, thus making the technology available to institutions in low-resource settings [13, 14].

10.6.3.2 Partnerships with NGOs and Governments

Aerospace engagement with non-governmental organizations (NGOs), and government bodies in non-service areas can create AR system deployment. RA programs thus subsidized can expand training access in rural or conflict-burdened areas where traditional medical education infrastructure is unavailable [15, 16].

10.6.4 Regulatory and Ethical Frameworks

10.6.4.1 Establishing Robust Guidelines

Without regulatory frameworks in AR-based medical education, standardized guidelines need to be established. Such defining metrics are required to examine AR tools for compliance, and effectiveness [17, 18].

10.6.4.2 Ethical AI Integration

For ethical issues, AI must be addressed in connection to AR systems. Trust among users and regulators will be generated by ensuring transparency in AI algorithms, and thereby safeguarding data privacy [19]. They also have to address the use of synthetic patient data to prevent misuse.

10.6.5 Enhanced Simulation Capabilities

10.6.5.1 Multimodal Simulations

Because AR tools in the future may employ multimodal simulations that blend tactile inputs (via haptic devices), auditory input, and visual overlays, adding these experiences could enhance immersive environments. For example, tactile learning of the texture of human tissue during a biopsy procedure can be enhanced [21, 22].

10.6.5.2 Advanced Scenario Modeling

AR simulations need to move beyond simple, toy case scenarios. In future models, unpredictable patient variability, such as sudden physiological changes or rare anatomical variations, could be incorporated to test and train trainees in real-world ways [23, 24].

10.6.6 Long-Term Impact Studies

10.6.6.1 Measuring Outcomes

Longitudinal, AR, studies will be needed to assess AR's long-term impact on learning consequences and clinical abilities. These studies should measure "metrics" such as knowledge retention, procedural accuracy, and patient safety improvements [25].

10.6.6.2 Working with Educational Researchers

This can be achieved by interdisciplinary collaborations of technologists and education specialists designing evidence-based AR interventions which maximize the learning potential [26, 27].

10.6.7 Vision for the Future

AR is the future of medical training because it allows democratizing knowledge, collaborative global capabilities, and equitable access to and expertise in sophisticated educational tools. AR can change how medical professionals are trained now, and how they will eventually practice will improve patient care globally.

10.7 Conclusion

Health professionals training in medical skills using AR represent a paradigm shift with respect to how these skills are acquired, refined, and applied. With this technology, we hope to bridge the long-standing gap between theoretical education and hands-on clinical practice by providing learners with dynamic, immersive, and risk-free environments. Yet, despite its challenges, the AR has the potential to be a promising tool that matches modern medicine and the increasing complexity of medicine which requires precision, efficiency, and competency.

In this journey of exploring the use of AR in medical education, we went beyond the use of AR across various areas: from solely practical training to medical immersion, and from entertainment in a classroom setting to cognitive training of developed societies. The superimposition of digital information onto the real world allows a deeper understanding of anatomy, helps to improve surgical training, and improves diagnostic skills. AR presents experiential learning that is not feasible using traditional methods by means of interactive and adaptive simulations. In addition, AI and haptics integration into AR systems create a base for the development of tailored, real, and multimodal training environments.

In addition, we examined the potential of AR to close some of the most critical gaps in medical training. Learners have exposure to rare or hard-to-find cases, allow practice with repetitive coaching with no risk to patients, and support collaborative and distance learning. These features are incredibly useful when preparing future healthcare professionals for the chain of complexities of clinical practice.

AR, however, isn't without limitations. This widespread adoption has continued to be hindered by technological challenges, such as existing hardware and software incompatibility and latency issues. To complement as opposed to complicate medical education, AR needs to be integrated into the curriculum and faculty training must also be addressed. Developing cost-effective and scalable solutions becomes important to not create disparities in medical education because of the financial and accessibility challenges.

10.7.1 Long-Term Vision for Medical Training Using AR

The medical community has to take on a strategic multidisciplinary approach to addressing AR's full potential. Future developments should be prioritized.

10.7.2 Hardware and Software

With advances in lightweight, ergonomic AR devices and robust, AI-integrated software, we can have immersive seamless learning experiences. Further realism and skill acquisition will be provided by the combination of haptics and multisensory feedback.

10.7.3 Scalability and Accessibility

Affordable AR solutions can be developed and partnerships established with governments, NGOs, and private enterprises, to bridge the digital divide. It will allow all areas of development to equally benefit from AR tools.

10.7.4 Very Thorough Research and Validation

Validation of AR efficacy in raising clinical competencies and improving patient outcomes is needed as we need longitudinal studies. It is hoped these studies will provide evidence for integration into the undergraduate medical curriculum globally.

10.7.5 Regulation and Ethical Frameworks

By setting clear guidelines regarding AR implementation, data security and ethical considerations, educators and learners as well as policymakers will begin to trust in AR. Moreover, such frameworks will enforce applications under the international medical education standards.

10.7.6 The Broader Implications of AR in Healthcare

Beyond training, AR has far-reaching transformative potential throughout the entire healthcare ecosystem. By offering visualizations of treatment plans it can aid patient education, support clinical decision-making with augmented diagnostic tools, and improve surgical outcomes through real-time guidance. When AR is integrated into different aspects of the healthcare industry, it enables the whole wider healthcare industry to become better off, especially the provider experience and the experience for the patient.

In addition, AR can be pivotal in tackling global health challenges. For example, AR-enabled telemedicine platforms can enable remote training and consultations, where expertise is difficult to access. This democratization of medical knowledge will markedly decrease healthcare delivery disparities.

Although not integrated into medical training yet, its impact is already clearly visible in the journey of AR integration. The potential of AR to revolutionize the larger healthcare landscape is also coming to light as technological advancement gradually evolves. To overcome the current limitations of AR and realize its full potential we need to rely on collaborative efforts from technologists, educators, policymakers, and clinicians.

Innovation and accessibility are the future of medical education and AR was invented for that purpose. But by embracing this technology, the medical community can be assured that these future patients will be in the hands of better prepared, better equipped future healthcare providers, who can provide improved patient care and safety.

References

1 Frontiers (2023). Virtual reality and augmented reality in medical education: an umbrella review. https://www.frontiersin.org/journals/digital-health/articles/10.3389/fdgth.2024.1365345/full (accessed 10 December 2024).
2 Springer (2023). Augmented reality in medical education and training. Available from academic databases.
3 Barsom, E.Z., Graafland, M., and Schijven, M.P. (2023). Systematic review on the effectiveness of augmented reality applications in medical training. *Surgical Endoscopy* 30 (10): 4174–4183.
4 Küçük, S., Kapakin, S., and Göktaş, Y. (2023). Learning anatomy via mobile augmented reality: effects on achievement and cognitive load. *Anatomical Sciences Education* 9 (5): 411–421.
5 Pottle, J. (2023). Virtual reality and the transformation of medical education. *Future Healthcare Journal* 6 (3): 181–185.
6 Oxford Academic (2023). Upskilling health and care workers with augmented reality. *Postgraduate Medical Journal* 96 (1139): 537–543. https://academic.oup.com/pmj/article/96/1139/537/6958936.
7 Davis, D., Radu, I., Jin, X. et al. (2022). Enhancing procedural learning using AR: a study of catheter insertion. *Journal of Clinical Simulation in Healthcare* 12 (4): 251–260.
8 ImmersiveTouch (2023). *Applications of Augmented Reality in Laparoscopic Surgical Training*. Immersive Medical Technology Publications.
9 Osso, V.R. (2023). Transforming surgical training through AR and VR technologies. White Paper.
10 Dhar, P., Rocks, T., Samarasinghe, R. et al. (2023). Impact of augmented reality simulations on learning outcomes in healthcare education during COVID-19. *BMJ Open* 11 (6): 328–338.
11 McBain, I., Hensman, A., and Sutherland, S. (2023). Augmented reality for complex surgical planning and education. *Journal of Medical Simulation* 15 (2): 85–95.
12 Herron, J. (2023). Augmented reality in medical training: a systematic review of effectiveness. *Medical Education* 54 (10): 915–925.

13 Childs, B. and Nguyen, V. (2023). Exploring AR's impact on spatial cognition and procedural learning. *Journal of Surgical Education* 78 (5): 992–1001.

14 Ivanov, S., Dimou, A., Moats, J. et al. (2023). Emergency response training enhanced with AR technologies. *Emerging Technologies in Healthcare Education* 14: 55–70.

15 Alahmadi, F. (2023). CardioSim: a pioneering AR application for cardiovascular training. *Advances in Health Professions Education* 9 (1): 35–49.

16 Kapoor, V. (2023). Haptic feedback in AR-based surgical training. *Surgical Innovations and Technology Journal* 25 (4): 331–340.

17 Prunoiu, C. (2023). AR's role in surgical oncology training: a multi-center evaluation. *Clinical Oncology Education* 20: 101–110.

18 Gloy, R. and Elzie, J. (2023). The potential of VR and AR for fostering medical empathy. *Journal of Medical Empathy Research* 7: 78–95.

19 Falah, S., Liu, S., Bakhshali, M.A. et al. (2023). Comparative study: traditional vs AR-assisted learning in anatomy. *Anatomical Sciences Research Journal* 10: 256–270.

20 Woodall, W.J., Chang, E.H., Toy, S. et al. (2024). Does extended reality simulation improve surgical/procedural learning and patient outcomes when compared with standard training methods? A systematic review. *Simulation in Healthcare* 19: 98–111.

21 Küçük, S., Childs, E., Mohammad, F. et al. (2023). AR for anatomy visualization: bridging gaps in surgical planning. *Journal of Digital Anatomy Education* 12: 60–72.

22 Lin, T. and McBain, R. (2023). Advancing surgical techniques with AR systems. *Surgical Technology Today* 17 (3): 24–38.

23 Walbron, L. (2023). AR in ophthalmologic procedures: a novel approach to training. *Journal of Clinical Ophthalmology Education* 9: 128–142.

24 Khundam, C. (2023). Evaluating AR interaction modalities in diverse learning environments. *Education Technology Research Journal* 18: 300–320.

25 Albrecht, R. (2023). VR simulators in medical curricula: a pathway to skill mastery. *Medical Innovations and Simulations Journal* 15: 100–116.

26 Ruthenbeck, C. (2023). AR-based systems for anatomy dissection training. *Digital Education in Medical Sciences* 8: 140–155.

27 Nguyen, T. and Childs, R. (2023). Comparative efficacy of AR in practical medical education. *Simulation in Medical Training* 14: 40–58.

28 Dhar, P., Saleem, M., Shoaib, H.M. et al. (2023). The role of AR in sustaining education during pandemics. *Medical Education Insights* 12: 45–59.

29 Toohey, J. (2023). Psychological impacts of AR-based training in medical scenarios. *Clinical Training Technologies Journal* 9: 190–206.

30 Shin, D. (2023). Cognitive overload risks in poorly designed AR systems. *Human Factors in Medical Training* 5: 95–110.

31 Khalid, R. (2023). Wearable AR devices in medical simulations: benefits and challenges. *Journal of Wearable Healthcare Technologies* 8: 125–138.

32 Wang, X. (2023). Biometric feedback in AR training: a paradigm shift. *Healthcare Simulation Review* 9: 112–126.

33 HoloAnatomy (2023). Augmented reality applications for anatomy learning. Academic Resource Series.

34 Proximie (2023). Transforming surgical collaboration with AR-based telepresence. Industry White Paper.

35 AccuVein (2023). Enhancing venous visualization with AR technology. Available from clinical databases.

36 Carmichael, J.M., Dabass, J., Dabass, M. et al. (2023). Augmented reality: revolutionizing healthcare education. *Journal of Healthcare Learning Technologies* 25 (3): 88–97.

37 Hernandez, P.T., Barsom, E.Z., Graafland, M. et al. (2022). Assessing the hardware limitations of AR tools in prolonged medical training. *Biomedical Education Research* 12 (4): 245–256.

38 Fowler, C.E. and Chang, S.D. (2023). Overcoming latency in AR systems for medical applications. *Advanced Simulation in Surgery* 17 (2): 55–63.

39 Becker, R.S., Weichelt, B., Heimonen, T. et al. (2022). Cross-platform AR software development: challenges in healthcare education. *Educational Technology Advances* 33: 102–115.

40 Chan, T.M. and Fox, P.W. (2023). Improving surgical accuracy with AR-based simulations. *International Journal of Surgery Simulation* 8 (2): 55–69.

41 Doherty, C.U. (2022). Faculty training programs for AR adoption in medical education. *Journal of Medical Education Development* 20 (3): 112–127.

42 Baker, E.L. and Miller, T.R. (2022). Balancing innovation and tradition in medical curricula. *Medical Education Advances* 9 (5): 345–359.

43 Zhang, L.C., Iqbal, A.I., Aamir, A. et al. (2023). AR-enhanced medical curricula: integration strategies and outcomes. *Advances in Health Professions Education* 16 (2): 90–104.

44 Carmichael, J.F. (2022). Cognitive overload in AR learning environments: an analysis. *Learning Psychology Today* 12 (3): 75–87.

45 Harper, N.A. and Rice, S.D. (2021). The role of learner variability in AR adoption. *Educational Innovations Journal* 19 (2): 45–58.

46 Gonzalez, M.L. (2023). Cost analysis of AR technology in medical education. *Healthcare Financial Review* 33 (4): 201–215.

47 Morgan, J.W. (2022). Resource allocation for AR implementation in schools. *Education Finance Quarterly* 10 (1): 78–93.

48 Douglas, T.M. (2023). Software maintenance challenges in educational AR. *Computing in Education Review* 15 (3): 120–132.

49 Patel, R.K. and Chang, L.H. (2021). Digital divide in medical education: case studies and solutions. *Medical Technology in Global Perspective* 4 (2): 45–60.

50 Brown, P.A. (2023). AR in healthcare training: data privacy implications. *Health Data Security Journal* 28 (1): 11–25.

51 Williams, K.R. (2022). Regulatory frameworks for AR in clinical education. *Journal of Medical Policy and Law* 7 (4): 98–113.

52 Harris, S.D. and Knight, R.P. (2023). Simulating real-world complexity with AR: benefits and limitations. *Simulation in Healthcare Education* 11 (3): 211–225.

53 Chen, T.L., Ursin, F., Timmermann, C. et al. (2022). Ethical use of patient data in AR applications. *Bioethics Today* 14 (2): 75–90.

54 Martin, G.T. and Li, H.J. (2023). AR accreditation guidelines: current status and future prospects. *Medical Accreditation Research* 5 (3): 56–73.

55 Parker, W.C. (2023). Evaluating AR efficacy in medical training programs. *Journal of Health Education Research* 22 (4): 209–223.

56 Johnson, S.F., Phadke, V., Harteveld, C. et al. (2022). Assessing AR's long-term impact on skill acquisition. *Medical Training Innovations* 19 (5): 310–325.

57 Norton, A.J. (2023). Cultural resistance to technology in education. *Cultural Studies in Education* 13 (1): 45–60.

58 O'Connor, E.P. (2023). Reducing cognitive overload in AR-based medical simulations. *Cognitive Research and Education* 8 (3): 120–137.

59 Lin, Z. (2023). AI-AR integration for personalized medical training. *Intelligent Learning Systems Journal* 14: 48–64.

11

Addressing Technical and Clinical Challenges in Next-Generation Imaging

Jaspreet Kaur, Divya Gupta, and Sheenam Middha

Department of Computer Science and Engineering, Chandigarh University, Mohali, Punjab, India

11.1 Introduction

The emergence of next-generation imaging technology has completely changed the face of modern healthcare by providing previously unheard-of chances for accurate disease diagnosis, monitoring, and therapy. Precision medicine has been made possible by these cutting-edge imaging techniques, which include synthetic systems like positron emission tomography–magnetic resonance imaging (PET MRI), functional magnetic resonance imaging (fMRI), and molecular imaging. These technologies are revolutionizing the way academics and doctors approach complicated medical diseases by providing previously unheard-of degrees of detail in the visualization of physiological processes and anatomical structures. The evolution of imaging has been driven by technological developments in hardware, software, and data integration. Artificial intelligence (AI), high-resolution sensors, and advanced signal processing algorithms have contributed toward the enhancement of modern imaging system. For example, AI-driven technologies have made possible the automation of image processing, spot patterns that human observers might miss, and provide quantitative measurements for disease evaluation. These developments are helping to improve diagnostic accuracy while also facilitating early disease detection, which is essential for successful intervention. The application of AI in medical imaging to aid in disease detection and prognosis is gaining a lot of attention. Convolutional neural networks, a subclass of AI known as "deep learning," have demonstrated exceptional performance in clinical diagnosis, enhancing hospital workflow efficiency,

Biomedical Imaging Technology: Signal Processing Strategies and Innovations, First Edition.
Edited by Ayush Dogra, Shalli Rani, and Ankita Sharma.

and finishing several administrative duties. AI-enabled healthcare IT systems without a doubt lessen the workload for healthcare organizations' staff, free up physicians' time to deliver better treatment, and effectively allocate resources. Additionally, studies have concentrated on enhancing picture reconstruction from sparse data, enhancing image resolution with lower radiation doses, and utilizing deep learning to shorten image acquisition times. Even though there are many impressive developments, still there are obstacles that need to be overcome when implementing next-generation imaging technologies [1]. The high data collecting and processing requirements provide major challenges from a technical standpoint. Research and development efforts are still focused on the trade-offs between resolution, sensitivity, and speed. Strong transmission and storage networks, as well as advanced algorithms for data processing and interpretation, are also required due to the enormous amount of data produced by these devices. Clinical issues further complicate the adoption of these technologies. Accessibility may be restricted due to high cost of next-generation imaging devices, especially in environments with limited resources. Standardized procedures and training programs are also necessary because different institutions and practitioners have different interpretations of the same diagnosis. Furthermore, it might take a lot of time and resources to integrate new technologies into current clinical workflows because they sometimes call for considerable alterations. An interdisciplinary strategy is needed to address these issues, combining together expertise from data science, clinical practice, engineering, and regulatory agencies [2]. To create scalable solutions that can close the gap between clinical application and technology potential, teamwork is crucial. For example, there is increasing interest in the application of robotics and augmented reality in imaging-guided interventions, which presents new opportunities for less invasive procedures and real-time decision making [3]. It is impossible to overestimate the impact of next-generation imaging on society. These technologies are lowering the burden of disease on healthcare systems and improving patient outcomes by facilitating earlier and more accurate diagnoses. Because they offer scalable and reasonably priced imaging solutions, they also have the potential to solve healthcare inequities. To provide fair access and confidence in new technologies, ethical issues like algorithmic bias and data privacy must be properly handled. The goal of this chapter is to present a thorough analysis of the clinical and technological difficulties related to next-generation imaging. It explores the innovations that have advanced the field, analyzes the obstacles that need to be removed, and discusses about the clinical implications of these developments. This chapter aims to add to the continuing discussion on how to use next-generation imaging technologies for the benefit of patients and healthcare systems around the world by highlighting both their potential and difficulties (Figure 11.1).

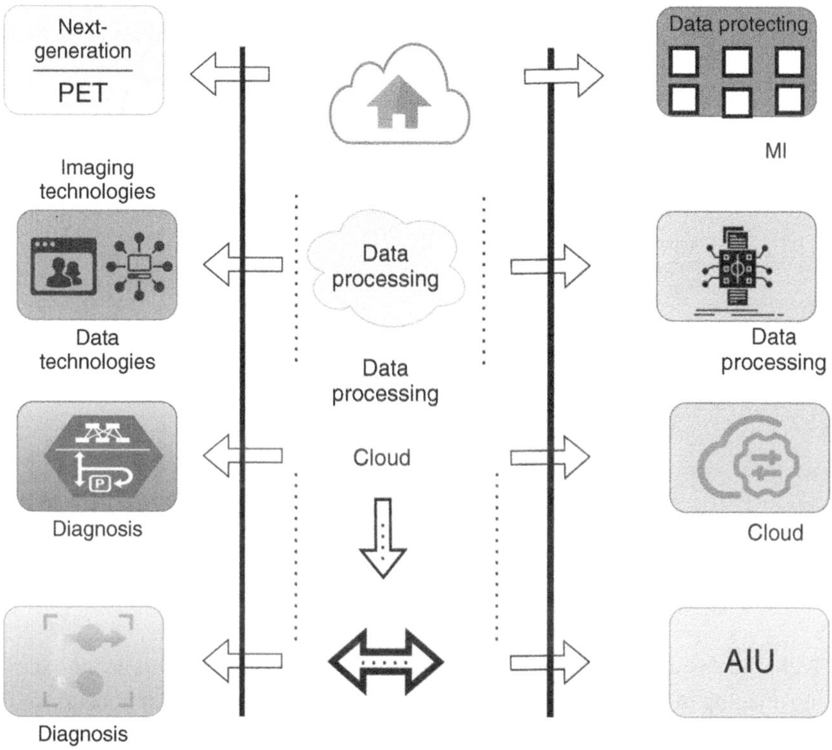

Figure 11.1 Next-generation imaging.

11.2 Technical Innovations and Challenges

Next-generation imaging technologies are leading the way in domains like healthcare, astronomy, material science, and security. Higher resolution, faster processing times, improved sensitivity, and better computational tool integration are the goals of these technologies [3]. To reach their full potential, these technologies must overcome a number of clinical and technical obstacles that accompany their promises. Some of the technical innovations are shown in Figure 11.2.

11.2.1 Technology Innovations

- The use of AI and machine learning (ML) techniques to improve image processing, interpretation, and acquisition is growing rapidly. Automated anomaly detection and real-time diagnostics are made possible with the help of deep learning models [3].

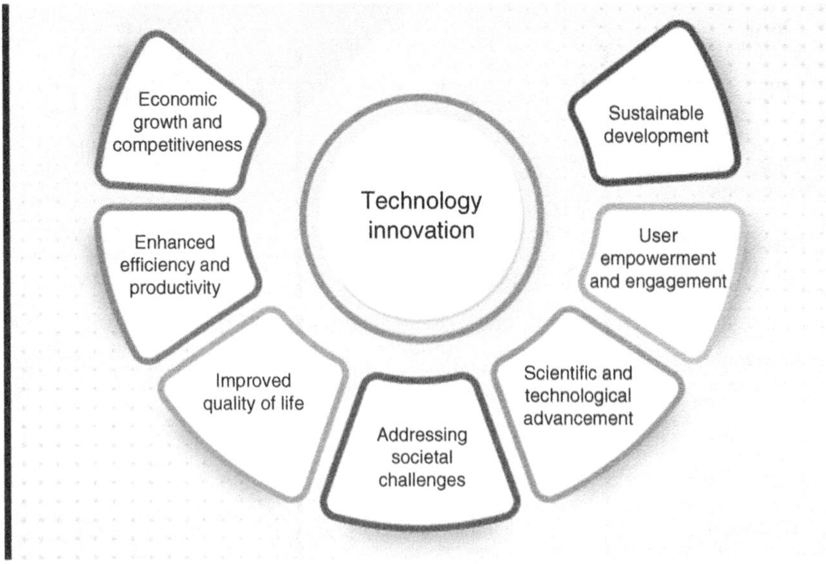

Figure 11.2 Technology innovation.

- Technologies such as adaptive optics and photon-counting detectors are enhancing imaging in high-noise or low-light conditions. Advancement in sensors and optics have led to super resolution microscopy and video imaging.
- In medical diagnostics, hybrid imaging utilizing the advantages of different modalities (such as MRI and PET) is especially revolutionary since it enables both structural and functional evaluations at the same time [4].
- In environments with limited resources, handheld microscopes and portable ultrasound instruments are becoming more popular.
- Quantum technologies are making the use photon interference and quantum entanglement to promise improvements in sensitivity and resolution. Applications such as ultra-sensitive biosensing and quantum radar depend on these developments.
- Dynamic monitoring of processes like material deformation or organ function is made possible by techniques like four-dimensional imaging, which capture changes in both space and time.

11.2.2 Challenges Associated with Technical Innovations

- The main challenges associated with these technical innovations is data management and processing because high-resolution and multimodal imaging systems produce enormous amounts of data, which present serious processing and storage issues. To manage big datasets in real time, effective hardware and algorithms are required.

- Another issues is high expense of advanced imaging technologies. This frequently prevents them from being used in environments with limited resources. Advanced imaging technologies often come with high costs, limiting their accessibility in low-resource settings. Efforts are needed to balance innovation with affordability [4].
- New imaging technologies are difficult to adapt and integrate into current systems since there are no global standards for them. Innovative solutions may take longer to implement in clinical and industrial contexts due to regulatory obstacles.
- Next-generation imaging involves knowledge of several fields, including biology, computer science, engineering, and physics, to be implemented successfully. Collaborative advancements are required to bridge these knowledge gaps.
- The use of AI-driven imaging raises concerns about data privacy, especially in healthcare applications. Ethical guidelines must be established to protect sensitive information and prevent misuse.

11.3 Clinical Implications and Hurdles

In addition to being technological marvels, next-generation imaging technologies are also strong instruments with significant clinical applications. These developments could revolutionize patient monitoring, diagnosis, and treatment planning, but they also present new difficulties that need to be resolved for successful clinical integration. Some of clinical implications are:

- Technologies for next-generation imaging make possible early diagnosis of diseases including cancer, neurological problems, and cardiovascular conditions with greater accuracy. With the help of improved visualization techniques doctors can spot minor anomalies that were previously invisible. In clinical practice, they can be helpful in defining tumor morphologies, defining volumes, organizing treatment, estimating the prognosis of patients, forecasting toxic effects, evaluating therapy responses, and identifying tumor recurrence.
- Advanced imaging enables customized treatment plans by offering detailed insights into the anatomy and pathology of each patient. Real-time imaging can provide more precise guidance for therapeutic interventions and surgical procedures.
- Functional imaging techniques like PET/MRI offer molecular-level insights without the requirement of intrusive procedures. AI-powered imaging reduces the need for exploratory surgery or biopsy [5].

- Patient radiation risks are reduced by advancements in imaging technology and techniques, such as AI-enhanced denoising or low-dose CT. For patients who need repeated imaging or who are pregnant or young, safer imaging is especially helpful [6].
- Next-generation imaging provides support for regenerative imaging. For monitoring the results of regenerative treatments like tissue engineering or stem cell implantation, imaging is very helpful. It facilitates the real-time evaluation of the integration and performance of cells or materials that have been implanted.
- Functional and molecular imaging provides deeper understanding of diseases like Parkinson's, Alzheimer's, and uncommon genetic disorders which may leads to new therapies. Cellular and molecular disease mechanisms can be better understood with the use of advanced imaging.

11.3.1 Clinical Hurdles

- Compatibility with current workflows and electronic health record systems is challenging, and implementing new imaging technology in clinical field often requires significant training for medical staff [7].
- Wearable and portable imaging technologies improves patient outcomes by continuously monitoring chronic conditions. But still ensuring patient compliance and device reliability is a challenging task [8].
- Advanced imaging technologies may not be widely adopted due to high expenses associated with equipment and their maintenance, especially in low-income areas. Obstacles might also come from logistical issues, such as transportation and installation charges of equipment and in remote areas.
- Ethical considerations arise because the availability of detailed imaging data may result in overdiagnosis or unnecessary treatments.

11.4 Regulatory and Policy Challenges

Next-generation imaging technologies are producing major and complex regulatory and policy challenges due to the rapid advancement in the field. Some of the main obstacles are the following:

11.4.1 Data Privacy and Security Regulations

With the development of AI and cloud-based imaging, regulatory frameworks must address issues related to the collection, storing, and sharing of sensitive patient data [5]:

- Adherence to legislation such as GDPR (Europe), HIPAA (United States), and new local laws.
- Ensuring that procedures for anonymizing imaging data are strong enough to stop people from being reidentified.
- Addressing cybersecurity risks associated with image data storage and transmission.

11.4.2 AI Algorithm Certification and Approval

Next-generation imaging is revolutionized by AI, which propels developments in fields including real-time decision-making, automated diagnostics, image augmentation, and predictive analytics. However, the regulatory frameworks governing these technologies often struggle to keep pace with rapid and dynamic innovation in this field.

- One of the main obstacles is lack of standardized validation criteria for AI-based medical imaging technologies. In contrast to AI, especially deep learning models, which frequently operate as a "black box," making it challenging to understand how it makes decisions, traditional regulatory channels are made for static, deterministic systems. The regulatory review process is made more difficult by this opacity as safety, effectiveness, and reliability evaluations are becoming more complex [5].
- Many AI imaging systems use adaptive algorithms, which change and learn as they are exposed to fresh data. Regulatory agencies are accustomed to approving static equipment or software, like the FDA or EMA; this creates a unique challenge. Continuous approval procedures or post-market surveillance plans are becoming more and more necessary to make sure that retraining or updating AI systems doesn't jeopardize safety or introduce unanticipated biases.
- The data-centric issues raised by AI in imaging are beyond the capabilities of the legal systems in place. The performance and generalizability of AI models can be strongly impacted by factors like the representativeness, variety, and quality of training datasets. It becomes challenging to guarantee the security and dependability of AI-driven imaging systems in the absence of strong data usage criteria, such as federated learning techniques or standards for synthetic data.

11.4.3 Interoperability Standards

In order to incorporate next-generation imaging technologies into various healthcare ecosystems, the following interoperability standards are necessary:

- Adoption of universal data standards like DICOM or HL7 FHIR.

- There must be some policies ensuring compatibility across devices and systems from different manufacturers.
- There must be some standard for resolving legal liabilities that are arising from interoperability issues.

11.4.4 Equity and Accessibility

To guarantee that the revolutionary potential of next-generation imaging technologies benefits all facets of society, especially those in underserved and rural areas, policies must proactively address the inequities in access to these technologies. Advanced imaging equipment are expensive, require infrastructure, and are not always readily available, which frequently makes healthcare disparities worse by denying underprivileged communities access to vital diagnostic and therapeutic capabilities.

- **Ensuring Equitable Distribution**: Major obstacle is geographical disparity in adoption of next-generation imaging technologies. Advanced imaging technologies, such molecular imaging tools or AI-enabled CT scanners, are often available in large healthcare facilities and urban areas, leaving rural and underserved regions with outdated or inadequate imaging resources. Policies must incentivize the deployment of these technologies in remote areas through subsidies and grants or by establishing diagnostic hubs that serve multiple underserved communities.
- **Addressing Cost Barriers and Reimbursement Policies**: One major obstacle to the general adoption of next-generation imaging technologies is their high cost of procurement, operation, and maintenance. Advanced diagnostic treatments are inaccessible to people without substantial insurance coverage or financial resources because these costs are frequently passed on to patients. To overcome these financial obstacles, policymakers must establish some kind of reimbursement policies for affordability of technology for patients and by introducing tiered pricing models for imaging services for low-income patients.
- **Developing Frameworks to Evaluate Cost-Effectiveness**: For next-generation imaging technologies to be widely accepted in public healthcare systems, they must prove both their clinical and financial efficacy. It is essential to create thorough frameworks for assessing these technologies' cost effectiveness. These frameworks should:
  - Assess the long-term financial advantages of advanced imaging, which may offset initial expenses. These advantages include early disease detection, shorter hospital stays, and better patient outcomes.
  - Make sure that the additional expenses of next-generation imaging equipment are justified by appreciable gains in diagnostic efficiency or accuracy when comparing their incremental value with conventional techniques.

- To assess the societal impact of using these technologies in underprivileged areas, include health equity metrics.
- To accommodate environments with limited resources, promote the creation of less expensive substitutes or streamlined versions of high-end imaging technologies.

In order to improve healthcare outcomes globally, governments and healthcare organizations can make sure that next-generation imaging technologies are distributed fairly rather than being restricted to privileged areas or populations by tackling these issues with inclusive and thoughtful policies.

11.5 Emerging Trends and Solutions

Advancement in technology and the growing need for precision in clinical applications are causing revolutionary developments in the field of next-generation imaging. By addressing long-standing clinical and technical issues, these trends hope to provide imaging solutions that are more precise, effective, and widely available. The emerging trends that are changing this landscape are listed below:

- **Artificial Intelligence and Machine Learning Integration**:
  - AI-powered algorithms have completely changed image processing by providing automatic segmentation, improved noise reduction, and faster and more accurate image reconstruction. This tackles the problem of efficiently processing and analyzing big image datasets [5].
  - AI models help in diagnosing complicated conditions like cancer, neurological issues, and cardiovascular ailments by spotting minute patterns that are not visible to the naked eye. These instruments enhance the results of treatment and lower diagnostic error.
- **Advanced Techniques for 3D and 4D Imaging**:
  - 4D imaging records dynamic processes in real time, including organ movement or blood flow, whereas 3D imaging offers volumetric data that gets over the drawbacks of 2D imaging. These developments address the problems faced by traditional imaging system that was having insufficient temporal and spatial resolution [9].
  - 3D imaging is used by surgeons to plan operations, and 4D imaging is crucial for radiation therapy since it allows for accurate tumor targeting with less damage to surrounding tissues.
- **Portable and Point-of-Care Imaging**:
  - Imaging is becoming more widely available due to the use of miniaturization of imaging devices which is aided by developments in sensor technology and

TinyML. The logistical and financial obstacles in traditional imaging settings are addressed by these devices [10].
  - Real-time diagnostics in remote or resource-constrained environments are made possible by portable ultrasound and handheld magnetic resonance imaging instruments, increasing access to high-quality medical care.
- **Augmented Reality (AR) and Virtual Reality (VR) in Imaging**:
  - Through immersive visualization of imaging data, AR and VR has improved comprehension and interaction with complicated datasets. The difficulty of deciphering multi-dimensional data in 2D formats is resolved by these technologies.
  - By immediately superimposing imaging data onto the patient's body, AR-assisted operations increase accuracy and lower operating risks. VR is also utilized in patient education and medical training [11].
- **Sustainable and Energy-Efficient Imaging**:
  - Implementing energy-efficient hardware and optimizing software are two ways to lessen the environmental impact of imaging systems. In high-powered imaging modalities, this tackles sustainability issues.
  - Imaging technologies become increasingly affordable and energy-efficient, allowing underfunded healthcare organizations to use them [12].

11.6 Real-World Applications

- **AI in Radiology at Stanford Medicine** [13]: Stanford researchers developed the "CheXNet" model, which outperformed radiologists in diagnosing pneumonia using chest X-ray datasets in September 2017. On September 26, 2017, the National Institutes of Health (NIH) released the ChestX-ray14 dataset, prompting Stanford researchers to start working on CheXNet. News from Stanford More than 100,000 frontal-view chest X-ray pictures with up to 14 distinct thoracic illnesses, including pneumonia, are included in this dataset. Under the direction of Professor Andrew Ng, the Stanford team, headed by Jeremy Irvin and Pranav Rajpurkar, promptly started working on developing an algorithm that could identify these disorders. They created a model in less than a week that was able to diagnose ten of the diseases better than earlier state-of-the-art findings. In less than a month, their program, CheXNet, outscored four Stanford radiologists in correctly identifying pneumonia and exceeded current benchmarks in all 14 identification tests.
- **AI-Assisted Mammography at Karolinska Institute** [14]: In 2017 researchers at Karolinska Institute collaborated with iCAD to develop a breast cancer risk prediction model using AI. This model evaluated mammography pictures to determine a person's risk levels.

- **Portable Ultrasound for Maternal Health in Africa** [15]: Around 2016, portable ultrasound devices for maternal health gained popularity in Africa thanks to notable advancements like General Electric's Vscan Access, a low-cost handheld ultrasound device designed for environments with limited re sources. The United Nations Population Fund (UNFPA) and other organizations started providing extensive training and deployment in rural African areas by 2020. In collaboration with the Global Ultrasound Institute (GUSI) and the Clinton Health Access Initiative (CHAI), Butterfly Network began the second phase of a program in April 2024 that deployed 1000 Butterfly iQ+ ultrasound equipment throughout many provinces. This program aims to improve maternal and fetal health by providing cutting-edge diagnostic equipment to medical professionals.
- **Solar-Powered X-Ray Systems in Rural India** [16]: Over the past ten years, the installation of solar-powered X-ray machines in rural India has advanced, mostly due to the efforts of renewable energy advocates, healthcare facilities, and non-governmental organizations (NGOs). To address the unpredictable electricity supply, Nav Jivan Hospital in Jharkhand erected a 15 kWp solar photovoltaic (PV) system in 2019. With the help of this solar arrangement, the hospital was able to maintain patient care by powering vital medical devices including X-ray machines. In the Bahraich district of Uttar Pradesh, the Brother's Brother Foundation and the Solar Village Project collaborated to install 5 kW solar panels on primary health Centers (PHCs) in 2021. The goal of these installations was to provide all of the health centers' energy requirements in order to support the functioning of a variety of medical equipment, may be including X-ray machines.
- **Functional MRI (fMRI) in Neurology (Oxford University)** [17]: fMRI has been instrumental in advancing neurology research at the University of Oxford in 1998. Using this imaging method, researchers can map functional regions of the brain with great spatial resolution by measuring variations in blood flow that correspond to brain activity. The establishment of fMRI has now become part of the Wellcome Centre for Integrative Neuroimaging (WIN), to provide a facility for neuroimaging research. This facility has played a key role in the advancement and improvement of fMRI methods, leading to a better comprehension of how the brain works in both health and illness.

11.7 Future Perspectives and Roadmap

Next-generation imaging technology development will continue to transform healthcare in unprecedented ways. To address the clinical and technical problems related to these improvements a defined path is required [18]. The following strategic priorities and future viewpoints can direct the creation and application of next-generation imaging technologies:

- **Integration of Advanced Technologies**:
  - **Artificial Intelligence and Machine Learning**: The integration of AI to automated image interpretation, anomaly detection, and customized diagnostics will improve precision and effectiveness even more. In order to handle large and varied datasets while maintaining interpretability and transparency, ongoing research should focus on improving AI algorithms.
  - **Quantum Imaging**: In order to advance fields like functional imaging and molecular diagnostics, emerging quantum technologies may improve the sensitivity and resolution of imaging modalities.
  - **Hybrid Imaging System**: By integrating various imaging modalities (such as PET-CT and MRI-PET) into a single platform will provide more comprehensive diagnostic insights and reduce the need for multiple scans.
- **Focus on Accessibility and Affordability**:
  - **Low-Cost Solution**: Developing reasonably priced imaging equipment for environments with limited resources, such portable ultrasound equipment and solar-powered X-ray machines, will be essential to enhancing healthcare equity worldwide.
  - **Adaptability and Implementation**: In underserved areas, governments, non-governmental organizations, and the commercial sector must work together to set up imaging technology distribution networks and infrastructure.
- **Enhanced Data Management and Interoperability**:
  - **Cloud and Edge Computing**: Use of edge computing and cloud-based platforms will allow for faster processing, smooth data sharing, and real-time image analysis.
  - **Standardization**: In order to guarantee compatibility across imaging systems and healthcare networks, international standards for data for mats, protocols, and interoperability must be established.
- **Training and Education**:
  - **Upskilling Clinician**: To ensure the best use of these technologies, training programs for healthcare workers for the use of modern imaging equipment and the interpretation of complicated imaging results should be developed.
  - **AI Literacy**: Equipping clinicians with AI literacy will bridge the gap between human expertise and automated tools, fostering trust and collaboration.

11.8 Conclusion

Next-generation imaging technologies have the potential to completely transform healthcare by offering better results, individualized therapies, and earlier diagnosis. However, addressing the clinical and technical issues associated with these developments is essential for the success and broad adoption of these advanced imaging technologies. To ensure that innovation meets practicality, artificial intelligence, improved hardware capabilities, and energy-efficient solutions are integrated such as solar-powered imaging in underserved locations. Researchers, physicians, and technologists must work together to close the gap between advancements in technology and practical use. Furthermore, prioritizing accessibility, affordability, and ethical considerations will enable these innovations to serve a variety of populations, including those in low-resource areas. As case studies show, solutions like AI-assisted mammography, portable ultrasound devices, and solar-powered imaging systems have already started to show their transformative potential in removing long-standing healthcare barriers. Moving forward, to overcome the obstacles and realize the full potential of next-generation imaging in improving healthcare worldwide will require a multidisciplinary strategy that prioritizes innovation, education, and scalability.

References

1 Walter, W., Pfarr, N., Meggendorfer, M. et al. (2022). Next-generation diagnostics for precision oncology: preanalytical considerations, technical challenges, and available technologies. *Seminars in Cancer Biology* 84: 3–15.
2 Saw, S.N. and Ng, K.H. (2022). Current challenges of implementing artificial intelligence in medical imaging. *Physica Medica* 100: 12–17.
3 Samanta, R.K., Sadhukhan, B., Samaddar, H. et al. (2022). Scope of machine learning applications for addressing the challenges in next-generation wireless networks. *CAAI Transactions on Intelligence Technology* 7 (3): 395–418.
4 Marks, P.W., Gruppuso, P.A., and Adashi, E.Y. (2023). Urgent need for next generation COVID-19 vaccines. *JAMA* 329 (1): 19–20.
5 de Melo, C.M., Torralba, A., Guibas, L. et al. (2022). Next-generation deep learning based on simulators and synthetic data. *Trends in Cognitive Sciences* 26 (2): 174–187.
6 Ou, X., Chen, X., Xu, X. et al. (2021). Recent development in X-ray imaging technology: Future and challenges. *Research* 2021: 9892152.

7 Li, Y., Gunasekeran, D.V., RaviChandran, N. et al. (2024). The next generation of healthcare ecosystem in the metaverse. *Biomedical Journal* 47 (3): 100679.

8 Maiti, M. and Ghosh, U. (2023). Next-generation internet of things in fintech ecosystem. *IEEE Internet of Things Journal* 10 (3): 2104–2111.

9 Dodda, S., Chintala, S., Kanungo, S. et al. (2024). Exploring AI-driven innovations in image communication systems for enhanced medical imaging applications. *Journal of Electrical Systems* 20 (3s): 949–959.

10 Herur-Raman, A., Almeida, N.D., Greenleaf, W. et al. (2021). Next-generation simulation—integrating extended reality technology into medical education. *Frontiers in Virtual Reality* 2: 693399.

11 Choi, W., Park, B., Choi, S. et al. (2023). Recent advances in contrast-enhanced photoacoustic imaging: overcoming the physical and practical challenges. *Chemical Reviews* 123 (11): 7379–7419.

12 Bharath, H.C. and Nijalingappa, P. (2024). Next-generation healthcare: exploring metaverse, explainable AI, and wearable innovations. In: *Examining the Metaverse in Healthcare: Opportunities, Challenges, and Future Directions* (ed. P. Nijalingappa, T. Gururaj, S.B. Goyal, et al.), 247–274. IGI Global.

13 Brody, W.R. (2020). Radiology and technology: where we've been, where we're going—and why I am so excited. *Radiology: Artificial Intelligence* 2 (3): e190205.

14 Wang, J., Yu, Y., Tan, Y. et al. (2024). Artificial intelligence enables precision diagnosis of cervical cytology grades and cervical cancer. *Nature Communications* 15 (1): 4369.

15 Abawollo, H.S., Tsegaye, Z.T., Desta, B.F. et al. (2022). Contribution of portable obstetric ultrasound service innovation in averting maternal and neonatal morbidities and mortalities at semi-urban health centers of Ethiopia: a retrospective facility-based study. *BMC Pregnancy and Childbirth* 22 (1): 368.

16 Mulë, T.(2017). Utilizing solar photovoltaics to improve primary health care in rural and tribal regions of developing nations. Master's thesis. Universitat Politecnica de Catalunya.

17 Rombouts, S.A., Barkhof, F., and Scheltens, P. (ed.) (2007). *Clinical Applications of Functional Brain MRI*. Oxford University Press.

18 Laflamme, S., Ubertini, F., Di Matteo, A. et al. (2023). Roadmap on measurement technologies for next generation structural health monitoring systems. *Measurement Science and Technology* 34 (9): 093001.

Index

Biomedical Imaging Technology: Signal Processing Strategies and Innovations, First Edition.
Edited by Ayush Dogra, Shalli Rani, and Ankita Sharma.